Living within Limits

A Scientific Search for Truth

Kenneth M. Merz, Sr.

Algora Publishing
New York

Library of Congress Cataloging-in-Publication Data —

Merz, Kenneth M., 1922-2007.
Living within limits : a scientific search for truth / Kenneth M. Merz.
p. cm.
Includes bibliographical references and index.
ISBN 978-0-87586-585-0 (trade paper: alk. paper) — ISBN 978-0-87586-586-7 (hard cover: alk. paper) — ISBN 978-0-87586-587-4 (ebook) 1. Truth. 2. Science—Philosophy. I. Title.

Q175.32.T78M47 2008
121—dc22

2008016761

Front Cover: Clockwise from left:

1. The "Face" on Mars

This aerial photograph of a Martian butte, taken by Viking in 1976, was seized upon by the paranormal community as evidence of civilization on Mars, even though such illusions are common in nature. © 1989 Roger Ressmeyer/NASA/CORBIS

2. Rapa Nui National Park, Rano Raraku volcano, Moai status © Massimo Ripani/ Grand Tour/Corbis

3. Model of Imperial Rome at the Time of Constantine: Area Around the Colosseum © Vanni Archive/CORBIS

Printed in the United States

Table of Contents

Part I Limits

Chapter 1. The Limits of Life

It is increasingly apparent that the earth has physical and biological limits. In particular, such physical resources as energy and arable land are limited. And as we shall see in the chapter, The Search for Truth, humans have severe limitations in both sensory and reasoning powers.

The human population is increasing; longevity is increasing; technological exploitation is increasing; and third-world scientific achievements in technology and expectations of improving living standards grow with each day. As a result, we are ever so slowly crossing a great divide that will someday be known either as the Great Awakening or the Sybarite Slumber. We are not yet fully awake to the fact that our consumption of natural resources is increasing faster than those resources can be renewed.

Limits to Growth

A number of authors have documented limits to the resources on Earth and the role these limits played in the collapse of ancient civilizations. A. Toynbee was among the first historians to describe in detail the breakdowns and disintegrations of civilizations (Toynbee, 1947). Since then Meadows, Tainter, and Diamond have analyzed the causes of the collapse of civilizations.

Dr. Meadows and her colleagues at MIT have produced computer models confirming that there are quantifiable limits to the resources of Earth (Meadows, D., et al 1972, 1992, 2004). The first model, published in 1972, was generated for the Club of Rome. The objective was to investigate the long-term consequences of growth in population, industrial capital, food production, pollution, and resource consumption. Their overall conclusions were that:

1. If these growth trends continue, the limits to growth will be reached sometime in the twenty-first century, resulting in a sudden and uncontrollable collapse in population and industrial capacity.

2. It is possible to alter these trends to achieve economic and ecological stability and long-term equilibrium.

3. The sooner we decide to alleviate (somehow) the problems of number one, the greater the chances of success.

The book created a furor. Twenty years later, in 1992, when *Beyond the Limits* was published, there was relatively little notice. Apparently, readers had made up their minds; it was old news even though the book forcefully stated that we were already well beyond the limits projected in 1972. And the third book in 2004 projected more deterioration in natural resource supplies.

But not everyone was convinced that the Earth has limitations or is losing resources. One book (Simon & Kahn, ed. 1984) had more than 25 contributors, all of whom were skeptical or unconcerned about a scarcity or limit in the supply of resources. As economists, their main arguments were taken from economic data, not from observations of the Earth's biology and geology. Simon (1981) states unequivocally that there are no limits on energy now or foreseeable for the future.

However, by now we can take the rising price of gasoline and energy as the first tangible evidence for the truth of the conclusions of *The Limits to Growth.* Evidently the "invisible hand of the market place" is going to come heavily into play as one of the first observed result of Earth's limits and the growing battle for greater equality of living conditions between the haves of the industrialized world and the have-nots of the third world.

The Collapse of Complex Societies

Historical studies of past societies clearly indicate that societies fail and collapse because of a relatively small number of common causes. Tainter (1988) lists a wide endless array of past societies that have faltered and collapsed: Western Chou, Harappan, Mesopotamian, Egyptian, Hittite, Minoan, Mycenaean, Roman, Olmec, Mayan, Chacoan, Huari, and on and on. He defines the meaning of complexity and collapse then reviews their variety. He lists the causes of collapse under eleven major themes, some of which overlap or are redundant.

1. Depletion of vital resources

2. Establishment of a new resource base
3. An insurmountable catastrophe
4. Insufficient or inappropriate response to change
5. Other complex societies
6. Intruders and invaders
7. Class conflict, elite incompetence and misbehavior
8. Social dysfunction
9. Mystical factors
10. Chance combinations of adverse events
11. Economic factors

Tainter describes much of the criticism of these ideas of how civilizations collapse in terms of the interpretation of observed facts versus the logic of the arguments for collapse. He maintains that critics argue, for example, over whether or not deforestation and erosion occurred rather than whether deforestation or soil erosion could logically be a cause of collapse. He then decides that "if the logic of an argument is faulty, a discussion of (factual) matters is largely unnecessary" and concludes that "existing explanations of collapse logically cannot account for it." This point of view is untenable; ignoring fact or "matters" is not logic, it is anarchy.

The first cause of collapse in the above list is depletion of vital resources. Tainter gives considerable space to this argument. But he never explicitly states who or what depletes resources. Of course, it is humans who deplete resources by consuming them. Societies typically grow larger and population is always a major part of that growth, particularly among primitive societies without knowledge of birth control techniques. Although Tainter often mentions, infers, or cites others who point to population growth, he never points to it as a definite cause of collapse. In fact, after considerable discussion, he abruptly dismisses depletion of vital resources as a factor in collapse.

The reason Tainter gives for his dismissal is that societies that routinely overcome problems will not "sit by and watch the encroaching weakness without taking corrective action" and "dealing with adverse environmental conditions may be one of the things that complex societies do best." Therefore, collapse cannot happen? Such speculative reasoning is not proof.

Since many of the civilizations under consideration grew for centuries before collapsing, it appears that whatever causes collapse is a slow accumulative process. Both resource depletion (#1) and diminishing economic returns (#11) can be slow accumulative processes that might also go largely unnoticed by the populace. Observation of long-term trends could suffer from cultural or generational amnesia. Poor written records or short lives could perhaps have kept our ancestors from noticing that there were fewer large trees than when they were young, for example.

Tainter finally chose "diminishing economic returns" as his explanation for collapse, largely because it could be applied to any society. As he said, it was global in its reach. This economic cause requires recognition of some specific definitions from economic theory dating back to T. Malthus (1798), A. Smith (1776), and D. Ricardo (1817) on factors involved in the production of services and goods.

1. Average product — the output per unit of input

2. Average cost — cost per unit of output

3. Marginal product — the increase in total output resulting from one increase in input

4. Marginal cost (or return) is the increase in total cost resulting from the production of one more unit of output. When the value of the additional output is less than the increase in input cost, it is a diminishing marginal return.

Using these definitions, Tainter explains that "continued investment in sociopolitical complexity reaches a point where the benefits for such investment begin to decline, at first gradually then with accelerated force" and "increased costs of sociopolitical evolution frequently reach a point of diminishing marginal returns." He explains further that "increased investments in complexity fail to yield proportionately increasing returns."

After originating (usually with some advantage in food production such as acquiring more land, inventing the plow, fertilizers, crop rotation, and irrigation or river flooding), an emerging civilization soon evolves sociopolitical complexes of military elites, priesthoods, aristocracies, and their entourages that produce little as they increase with a steadily diminishing marginal return.

Modern economists, in fact, have found that diminishing marginal returns apply to every aspect of human activity including, agriculture, manufacturing, education, information processing, scientific research, technology, and military conquests. One of the examples Tainter cites is the long series of military conquests by the Romans around the Mediterranean world until they ran out of societies to conquer and plunder.

Economic data is available for ancient societies such as the Roman, which had a coinage system, written records, and extensive trade relations, and which regularly stole the treasuries and wealth of conquered nations. But primitive societies often produce little more than bones and stones from which we might measure value. From these, marginal return can only be inferred.

Collapse

In 2005, J. Diamond published the book *Collapse*, which attempted to explain the same phenomena but differed somewhat from Tainter's analysis. Diamond listed five main causes for the collapse of civilizations:

1. Environmental damage societies inadvertently inflict on their environment.

2. Climate change due to volcanic eruptions, the intensity of solar energy reaching the earth depending on Earth's axis of rotation, and the advance and retreat of the ice caps

3. Hostile neighbors; intermittent or chronic warfare

4. Decreasing support from friendly neighbors and trading partners

5. The society is unable to recognize its environmental problems early enough or to respond effectively to them.

Numbers 1 and 5, which are similar, generally fall into seven categories: deforestation, soil fertility losses, water management problems, over hunting, over fishing, introduced species, and human population growth with increased per capita impact.

Of the five major factors and seven categories Diamond proposes, the impact of excessive population growth can be inferred to be the most fundamental and pervasive. Diamond defines collapse as: "By collapse I mean a drastic decline in human population size and/or political/economic/ social complexity, over a considerable area, for an extended time." The list of collapsed societies Diamond describes is: Anasazi, Cahokia, Mayan cities, Moche, Tikopia, Tiwanaku, Mycenaean Greece, Minoan Crete, Great Zimbabwe, Harappan, Rome, and Easter Island.

In the case of the modern world, events are somewhat different but there are still overall limits. First, the human population, now over 6.5 billion, is still growing. Although the rate of growth is slowing, population is projected to reach 9 billion by about 2050. Second, longevity is increasing — people live longer due to medical advances and better food distribution. Finally, third world living standards, scientific capabilities, and expectations are rising. The latter two factors are a direct result of the diffusion of Western science and technology throughout the world. In addition, Diamond points out there is no possibility that the third world standard of living can rise to that of the industrial US, Canada, and Western Europe without abruptly and disastrously consuming the total available nonrenewable resources of Earth. But we are now witnessing this exact slow and uneven increase in worldwide living standards as typified by the advancements in China and India. Fortunately, the Chinese are trying to limit their families to one child per couple.

Diamond describes a number of relatively more primitive societies including Easter Island, Pitcairn Islands, Chaco Canyon, Mayans, Norse Greenland, Haiti, Dominican Republic, and Australia, as well as the giant China.

DAWN TO DUSK ON EASTER ISLAND

Easter Island is singled out for a very meticulous analysis of its geology, archeology, and historical evolution. What follows is a summary of Diamond's findings about the island and its civilization. Chapter 2, "Twilight at Easter," is a gem and should be read by everyone interested in the advances in archeological techniques over the last fifty years.

Easter Island is 66 square miles in area, 2300 miles west of the Chilean coast and 1300 miles east of the Pitcairn Island group. It is completely isolated and represents the most easterly extension of the seafaring Polynesian peoples from their origins in Indonesia.

The island was formed by three volcanoes erupting from the sea over the last million years. About 200,000 years ago the youngest volcano, Terevaka, in the northern corner of the triangle formed by the three, erupted covering 95 percent of the island with lava. Over the centuries this lava formed a rich soil that supported a dense tropical forest of giant palm trees. That is how the first Polynesians found it.

The terrain between the three volcanoes is gently rolling with no deep valleys. The beaches fall away sharply and provide little support for coral reefs or the fish and shell fish that normally inhabit them. The climate is subtropical; the island is about as far south of the equator as Miami is north of it.

The first Polynesians arrived on Easter about AD 900 or perhaps AD 700 in a few ocean-going log dugout canoes. According to oral tradition, the leader of the first expedition to arrive was Hotu Matu'u with his wife, six sons, and their families. They also had chickens and probably a few rats because bones of these first appeared in garbage heaps about that time. Chickens were the only domesticated animal; there were no pigs, dogs, or goats. The common Polynesian domesticated plants such as bananas, taro, sweet potatoes, sugarcane, and mulberry also appeared with the first settlers.

Forests with giant palms up to nine feet in diameter provided plenty of wood for dugout canoes, firewood, and building materials. There is no evidence that other

Polynesians ever came to trade or settle. Hotu Matu'u and his clan were alone, totally alone at the end of the world.

From the evidence of their activities, the population rapidly increased reaching a peak of about 15,000 by AD 1400. The main period of deforestation was from AD 900 to 1300. As the forest was cut down, the land was more intensively cultivated for food. A vast system of rock gardening was instituted, which involved placing large rocks in spaced arrangements and planting the crops in between them. The rocks conserved moisture, reduced rain erosion, and kept the soil warmer at night. Low walls were also built to reduce the drying tendency of the incessant winds over the relatively flat island. In fact, Easter Islanders would be noted among primitive societies for their rock gardens if they had not also carved the huge statues that are their worldwide signature.

What can be said of the Moai, the giant legless torsos with sharp chiseled heads that faced inland along the coastlines of Easter? What was their purpose? What fever drove the islanders to carve them? There were 393 Moai erected on 300 slightly raised stone ceremonial platforms that are a common feature of Polynesian culture. There are another 97 broken and abandoned along the roads used to transport the Moai. In addition, there are a number in various stages of completion still left in the quarries. The total number is 887. The statues were carved from about 1000 AD to 1620 when, according to oral tradition, the last Moai was erected.

The Moai were typically 13 feet tall weighing 10 tons with some reaching 32 feet and 75 tons. They were carved from the soft lava-based stone using harder obsidian stone chisels. The islanders had no metal. Moai were carved prone with the face pointed skyward. When the statue was about complete, the underlying stone was chipped away by the carvers from a trench along the side of the prone figure. It was then slid onto a wooden sled and transported to the planned site for erection. The best explanation is that the sled was dragged along a track made of parallel wooden rails held in place with crossbars and 50 or more islanders pulling on ropes to supply the motive power. At the erection site the Moai were slowly leveraged upright into their final positions. The rail system probably explains the disappearance of the original palm forests.

The zenith of Easter's civilization was centered on the period AD 1400 to 1600. By that time the rock gardens and Moai were largely in place and the forests gone. The society had become organized into 12 clan-based regions, each beginning on the coasts with their row of Moai facing inland and extending into the interior of the island like slices of a great pie. The clans were headed by local chiefs with an overall chieftain. Clans competed with each other to build more and larger Moai to honor their most important citizens. Everything seemed well organized and progressing normally.

But Moai production was beginning to decline; porpoises could no longer be hunted because there were no large palm trees left to build ocean going dugouts, and without wood, cremation could no longer be practiced. For cooking islanders were reduced to burning grass and crop waste.

Even palm nuts disappeared from the diet while rat bones increased in the garbage heaps. Without forest coverage, soil erosion increased. Indications of tension and disarray were coming from all directions. Signs of collapse were surfacing.

Finally, in 1680, the priests, clan chiefs, and chieftain were overthrown by internal rebels; civil wars broke out. People were buried where they fell. Collapse was in full swing. Then on Easter day April 5, 1722, the island was discovered by the Dutch explorer Jacob Roggeveen. Further contact with Europeans led to repeated epidemics of smallpox and other diseases that decimated the local populace as it had done earlier throughout the New World. In 1774 Captain Cook spent four days at Easter Island and noted broken and overturned Moai. The islanders were destroying not only their neighbors but also their special glory. The year 1838 recorded the last European mention of an erect statue. In 1862–3 Peruvian slave ships carried off 1500 islanders, reputed to be half the total population at that time. By 1868, no Moai was left standing. Missionaries counted only 111 islanders in 1872. Easter Island had come full circle from the landing of Hotu Matu'u and his 100 or so settlers in AD 900.

There appears to be little real difference between Tainter's emphasis on diminishing marginal return and Diamond's emphasis on limited natural resources for an isolated population. Tainter's Romans were literate, traders, and conquerors. Diamond's Easter Islanders were illiterate, without trading partners, and had limited natural resources. Both cultures were aggressively active against the changing conditions they perceived to be facing them. Rome collapsed and disintegrated into many unstable regions and finally the Dark Ages. The Easter islanders collapsed into what was essentially extinction. In the examples of collapse the role of excessive population growth was not accented by Diamond. However, Easter Island is a clear example of collapse due to population growth beyond the limited resources of a restricted area.

It is noteworthy that although historians seem reluctant to assign blame, they always describe in great detail the suffering, violence, and famine that decimates populations after civilizations collapse. Some regions of the Roman Empire were described as devoid of population after it collapsed. Could excessive population growth normally be the cause of final collapse?

Limits of a Finite Earth

When we look at our Earth, we find limits in every direction. The Earth has a limited diameter, circumference, land area, water, atmosphere, and on and on. Now, in the last few years, we have learned graphically (with dollar signs) that Earth is truly finite; it has a large but limited supply of oil, gas, coal, and uranium, for instance.

Where did fossil fuel come from in the first place? Unknown to us, long before humans evolved, it was created by living creatures (single celled photosynthetic bacteria) over a period of 1.5 billion years from about 1.8 to 3.3 billion years ago. These bacteria used the chlorophyll molecule to convert carbon dioxide and water into carbohydrate and oxygen or, as a chemist would put it:

$$CO_2 + 2H_2O \rightarrow [CH_2O]\ H_2O + O_2$$

The atoms in brackets stand for the atomic ratio of generalized glucose or carbohydrate compounds that almost all living creatures consume for growth, energy and life. The reaction is called the "water splitting reaction" because the free oxygen comes from the water molecules not the carbon dioxide (Atkins, 1991). Later, over hundreds of millions of years, trillions of creatures that consumed carbohydrates were fossilized under pressure into the form of oil, gas, and coal, all mostly hydrogen–carbon compounds.

Fossil Fuel Limits

The limit to the supply of fossil fuel oil was definitively defined by M. King Hubbert in 1956. Dr. Hubbert was trained in geology at the University of Chicago and during the 1930s taught geophysics at Columbia University. After World War II, he directed the Shell Oil Companies' research laboratory in Houston, Texas (Heinberg, 2005). After a long career studying the geology of commercial oil fields, he published his conclusions in 1949 that oil production would soon reach a peak and then begin a long decline. In 1956, he predicted that oil production in the United States would peak between 1966 and 1972. Economists and politicians dismissed his scientific reasoning and conclusions with snide remarks but the peak in United States fossil oil production occurred in 1970 anyway.

Scientists had predicted a limit to fossil fuel availability since the 1920s, but no one had accurately predicted a specific decline fourteen years before it actually happened. In addition, Dr. Hubbert described a mechanism that chronicled the birth, life, and death of an oil field. He found that after discovery, oil fields rapidly increase in output as more wells are drilled and the area of the field mapped. Once the field limits were known, well drilling continued between the original wells until a point of diminishing returns was reached, then the total daily production began to decrease. At that point of decrease it was found that half the total recoverable oil in the field had been extracted. This was a signal that half the oil was extracted but also that half was still recoverable. Oil field production per year follows a general hump backed bell curve. It is also noteworthy that anywhere from 30 to 50 percent of the original total oil is left behind dispersed in the sand and rock. Future research may unearth better methods to extract the oil from that residue.

Oil Sands, Oil Shale, and Coal Conversion

Wells are not the only way to recover oil from the earth. There are large deposits of oil sands in Canada and Venezuela. However, recovery from oil sands is a much more expensive task than from an oil well. Estimates run from an extra ten to twenty dollars per barrel (42 gallons) more than the cost of a barrel of oil from wells. That amounts to $0.24 to $0.48 per gallon more, before other refinery and distribution costs are summed up. Nevertheless, oil sands will be producing oil for world needs for a long time to come, but at steadily increasing prices.

Coal is our most plentiful fossil fuel. In 1996, it was estimated that there were about 1 trillion tons of accessible world coal reserves, but it is very polluting to mine and to use. After adjusting for world population growth and compensating for conversion to liquid fuel, in 2003 G. Vaux estimated coal usage in the US would peak in about 2032 and completely disappear by 2267. A "peak" does not mean an end but the beginning of a decline in production and an increase in price, which in this case would continue for about 230 years. To convert coal to liquid fuel suitable for cars, trains, and planes will also be a mammoth, costly undertaking amounting to an additional 40-45 dollars per barrel or $0.95 to $1.07 per gallon. But converting coal to oil has one advantage; it is a well known technology having been used commercially in South Africa by a company called SASOL Technology Ltd.

There is also reported to be 800 billion barrels of recoverable oil locked in the sedimentary rocks under the states of Colorado, Utah, and Wyoming. In the 1970s, the Exxon oil company spent over 5 billion dollars in an attempt to extract some of that oil profitably. The effort was a total failure. On May 2, 1982, now known as "Black Sunday," Exxon pulled out, causing a catastrophe among the local Coloradoans.

Shell and other oil companies are considering a process that would use electric heaters to heat the oil shale 2,000 feet underground for several years to soften it. The whole idea sounds like a flight of fancy. Shell is planning to make a final decision on the process by 2010. An estimated increase in cost due to this process is $55–70 per barrel or $1.31 to $1.62 per US gallon.

The oil sands of Canada are already producing over a million barrels of oil per day; coal conversion to diesel will become commercially feasible at present crude prices; but shale oil seems to be a dark horse. Summing up, oil will steadily increase in price but the final limit for oil use is still sometime in the future, perhaps as much as one or two centuries. That doesn't mean we can sit back and let the invisible hand of the market take care of the problem. The future is already upon us and will take care of nothing unless we lay the proper foundations today.

Alternate Energy Sources

There are a large number of alternate energy possibilities for research and development efforts. These include: biomass, wind power, nuclear fission and fusion, hydrogen, solar cells, and conservation. All of these methods will be commercially evaluated in this century. They all require huge amounts of technical development for commercialization and capital if put into production.

A fanciful favorite is to use the hydrogen in water to power fuel cells. It is said that we have plenty of water. Unfortunately, the laws of thermodynamics will have nothing to do with that idea. To get a mole of hydrogen gas from water you have to supply 68,400 calories of energy from somewhere, electrolysis of water for instance, which would require the consumption of some other fuel. When you burn hydrogen and atmospheric oxygen in a fuel cell to power an auto, you get back 68,400 calories of energy per mole of hydrogen. That is a total gain of zero and when handling costs for high pressure pipelines, factories, tankers, and service facilities are included a huge net loss.

Ethyl alcohol has been touted (by the George W. Bush Administration) as an alternate energy source for cars and trucks. Obviously, ethanol will burn as a fuel but if we burn alcohol, which comes from sources that can also be used as food sources for humans and livestock, how many humans are we willing to consign to death from starvation in the future? With a still growing population projected to reach nine billion by 2050, will they be fed with corn and sugarcane from the limited supply of arable land or provided with the fuel to drive their SUVs?

Nuclear fission plants will be built in the hundreds or thousands until the limited supply of uranium and places to store waste products become scarce. At least nuclear plants don't spew out noxious gases that we know of yet.

Atomic Fusion

The only process that might yield large amounts of energy indefinitely considering our present level of knowledge is atomic fusion not fission. But after 50 years of research (largely to find a material or method that can contain the process) and pilot plant development, the best it has produced is close to breakeven; it has not yielded more energy than it consumes. Most authorities believe commercial development is still fifty years in the future.

Fusion requires heating light atomic nuclei and forcing them together until they combine to form a single new nucleus. The most widely used materials are deuterium (D) and tritium (T), heavy isotopes of hydrogen which are available in seawater or can be formed from lithium. That is the easy part; D and T must then be heated to over 100 million degrees centigrade, which is the temperature that powers the suns in their nucleosynthetic cores. D and T would then be in the form of a gas like plasma that must be confined and compressed until fusion occurs and releases some energy in accord with Einstein's equation, $E=MC^2$. The energy released could then be harnessed for use in other applications. At present, containment is carried out in a Tokamak, a huge torus or doughnut shaped magnetic chamber. Since the plasma of nuclei and electrons consists essentially of positive and negative particles, it is somewhat controllable by strong magnetic fields.

However, the central problem is still how to contain a temperature of 100 million degrees centigrade or more somewhere here on Earth. Common furnaces to melt glass or steel or fire ceramics are limited to such materials as aluminum oxide, which melts at 2050° C. Induction furnaces can go a few thousand degrees higher but 100 million degrees for any length of time is out of reach with current technology.

As far as is known, no fusion apparatus has yet reached the stage of Ignition, the stage where more energy is produced than is consumed by the apparatus. Until that point is experimentally demonstrated, the commercial application of fusion to produce useful energy on Earth is still 25–50 years in the future.

Global Warming

Up to this point the main worry has been about getting enough fossil fuel to burn. But burning itself presents problems, huge problems. Burning any of the fossil fuels, coal, oil, or gas produces equivalent amounts of carbon dioxide:

$$\text{carbon} + \text{oxygen} = \text{carbon dioxide, or } C + O_2 \rightarrow CO_2$$

This equation does not include the varying amounts of hydrogen chemically bonded with carbon in fossil fuels. Factoring in the hydrogen would simply result in the additional reaction product of water molecules, H_2O.

Increasing amounts of carbon dioxide in the atmosphere has an almost imperceptible but startling effect on global weather. A blanket of carbon dioxide gas in the atmosphere acts to produce what has been likened to a "greenhouse effect"; radiation from the sun is absorbed by the earth but for a variety of causes, the earth reemits less radiation back into space than it absorbed. The result is that the earth warms very slightly over long periods of time. C. Keeling measured increasing CO_2 concentrations from the summit of Hawaii's Mauna Loa for almost 50 years. It is known that the concentration of CO_2 in the atmosphere has increased 32% in the last 150 years. But increasing CO_2 concentration has been difficult to correlate with global warming until recently.

A pioneering investigator in global warming over the last 25 years has been Jim Hansen, who is now director of NASA's climate research center at the Goddard Institute. The Reagan and two Bush administrations have tried to silence Dr. Hansen's warnings of impending increases in global temperatures repeatedly. Nevertheless, at a talk in December 2005 Dr. Hansen told his audience that the evidence for global warming had become overwhelming by the year 2000 and he discussed a recent graph produced by a Russian team working at the Vostok station in the Antarctica (Bowen et al, 2006).

The Russians took a boring through the polar ice cap at Vostok that was 3.6 kilometers or 2.24 miles long. Air bubbles in the ice made it possible to measure the concentration of CO_2 in the atmosphere at the time the bubble formed. And the ratio of the isotope deuterium to hydrogen in the H_2O molecules of the ice gave the temperature of the atmosphere at the time the ice formed. This made it possible to determine the carbon dioxide concentration and temperature over the past 420,000 years of the earth's history. Graphs representing CO_2 concentration and temperature tracked each other, forming peaks at 325,000, 240,000, and 125,000 years ago. Lows formed 350,000, 250,000, and 150,000 years ago.

In addition, geological studies of sea levels from geological coastal erosion measurements also track the oscillations of temperature and CO_2 concentration. All three measurements rise and fall in tandem. The implication is that when the first two variables reach a peak, sea levels will also rise and inundate coastal areas.

Finally, the Vostok data indicate a cyclic low about 20,000 years ago and a sharp increase in the last few hundred years (around the beginning of the industrial revolution) that go above all the peaks of the previous 420,000 years. The CO_2 level has never exceeded 300 parts per million; it is now 377. Average earth temperature has never exceeded 15.5 degrees centigrade; it is now 14.55 degrees, having increased 0.8 degree centigrade in the last 100 years. Dr. Hansen expects a 2 to 3 degree rise this century. Historically, a rise of that magnitude would cause the seas to rise about 15 to 35 meters or 45 to 100 feet; resulting in flooded coastal cities and arable lands.

Conclusions

The overall conclusion is that there is no energy producing process that is as cheap as drilling oil wells; nor is there a fuel that is as adaptable for easy transportation needs as oil. Undoubtedly every possible energy process will undergo research and development in the coming years; they will all be more costly and impose an increasing capital cost burden on our energy requirements. That is certain to be the result for windmills, wave action, photovoltaic cells etc. etc. We are almost imperceptibly, moving into a dramatically new environment for human energy needs. And finally, the brief discussion of global warming warns that however we live, we must use less energy; conservation is a necessity.

Bibliography

Alberts, B. et al., 1998, *Essential Cell Biology*

Atkins, P. W., 1991, *Atoms, Electrons, and Change*

Bowen, M. et al., 2006, *Technology Review*, Jul/Aug

Chang, R. 1988, *Chemistry*

Dennett, D. C., 1995, *Darwin's Dangerous Idea*

Diamond, J., 2005, *Collapse*

Diamond, J., 1997, *Guns, Germs, and Steel*

Gould, S. J.1989, *Wonderful Life*

Grants, B. R. & P. R., 1989, *Evolutionary Dynamics of a Natural Population*

Heilbroner, R.L. 1953, *The Worldly Philosophers*

Heinberg, R. 2005, 2nd. Ed., *The Party's Over*

Hubbert, M. K. 1969, *Resources and Man*

Kennedy, J., 1987, *Rise and fall of the Great Powers*

Malthus, T., 1798, *Essay on the principle of Population*

Mayr, E., 1991, *One Long Argument*

Meadows, D. et al., 1972, *Limits to Growth*

——, 1992, *Beyond the Limits*

——, 2004, *Limits to Growth: The 30-Year Update*

Noble, P., Deedy, J., 1972, *The Complete Ecology Fact Book*

New York Times, 5/15/06 "Computer Scientist's Fall"

Ricardo, D., 1817, *Principles of Political Economy*

Rifkin, J., 1985, *Entropy*

Schopf, J.W., 1999, *Cradle of Life*

Simon, J. L., 1981, *The Ultimate Resource*

——, 1996, *The Ultimate Resource 2*

—— and Kahn, H., Eds., 1984, *Resourceful Earth*

Smith, A., 1776, *The Wealth of Nations*

Strickberger, M. W., 2000, *Evolution*

Tainter, J. A., 1988, *Collapse of Complex Societies*

Toynbee, A. 1947, *A Study of History* (Somervell Abridgement)

Chapter 2. Living within Limits

Living creatures have limits in all dimensions, in particular lifespan. The bristlecone pines of the southwest United States are among the oldest life forms that have ever lived. They live in high, dry, hostile environments, seemingly less alive than dead. By taking core borings through the living and dead portions of their trunks, scientists have established ages up to 9000 years. The core borings show unique patterns of ring width depending on the amount of moisture during each year of growth. Comparing ring patterns that overlap from different trees indicates how they relate to each other and to specific dates.

A seemingly brief lifespan is also important to us but the limitations of memory, intelligence, and senses are sometimes beyond bearing. Other species have all our limitations and in addition lack consciousness.

The evolution of life is a long passage through innumerable concurrent changes in both the internal and external environments that envelop every living being. The internal environment consists of atomic and molecular activity that continually repairs and renews the DNA and structure of each living cell. The external environment is the physical and biological conditions that surround us each moment. These two operate independently but in combination produce the selection part of evolutionary natural selection.

The Limits of Population

There is a quotation from the introduction to Darwin's *Origin of Species* (1859) which begins: "As many more individuals of each species are born than can possibly survive...." In these few words Darwin defined one of the major observable characteristics of living creatures on Earth. Species will continue to exist only if individual adult members produce more young "than can possibly survive." Darwin put that observation to use in supporting his natural selection concept.

Perhaps the best demonstration of the truth of Darwin's quotation is the winemaking art in which simple one-celled creatures increase beyond the possibility of survival.

The Winemakers

Winemaking is a very ancient art. Long before humans arrived, wine was part of the life cycle of many species. Today there are at least two families of winemakers. *Homo sapiens* are the second; we make wine to drink and enjoy. The first is *Saccharomyces cerevisiae*, the species Ellipsoideus (we'll call it Saccharo E for short). This particular species is classified under fungi and is one of the yeasts that can carry out fermentation. Saccharo E make wine to live; they eat sugar and convert it to ethyl alcohol and carbon dioxide. These are very tiny creatures that have only one cell per individual and reproduce by budding or spores. Under optimum conditions they can double their numbers in an hour. But as we shall see it is a very dangerous life they lead. Saccharo E did not set out to live on sugar alone; they can ferment many kinds of organic materials. We found they made the best wine according to our taste and decided that was their real purpose.

Homo sapiens has been making and drinking wine for several millennia, apparently enjoying every minute of it. The winemaking process has been the subject of daily discussion and argument among devotees of the art for centuries. Those who drink wine never cease extolling their particular favorites. The winemaking process is simple enough for anyone who can follow a recipe. It consists chiefly of putting fruit juice, sugar, and Saccharo E together for about three months. The juice of grapes is the favorite, although everything from apples to dandelions to elderberries seems to do as long as the proper amount of sugar is included.

The modern method of winemaking begins by first cleaning and crushing fruit in a crock. This mess is treated chemically to kill all wild bacteria and fungi and adjusted to a concentration of about 22% sugar. A small amount of Saccharo E is added. There is a short initiation period; then in a few days of reproduction, a spectacular eruption of foam and bubbles occurs as Saccharo E attains its maximum of carbon dioxide gas production. It is a particularly special sight if your crock is not large enough; potential wine is all over the floor and you must scramble for additional containers. After this first strong fermentation, the juice is pressed from the seeds, skins, and stems then stored in jugs with water locks that prevent wild fungi from entering. By now, the bubbling is slowing and sediment settles to the bottom of the jugs. Over the next three months the clearing wine is decanted two or three times into clean jugs or barrels leaving the sediment behind. Finally the wine is bottled and aged. If this is your first batch of wine, aging will take about a week, but most authorities suggest a few months or a year.

The above is a very brief recounting of the winemaking art. There are endless variations and nuances at every step of the process that keep devotees searching for the perfect wine. Since the palate is the measure, we each have a different perfection.

The viewpoints of Saccharo E are quite different. Floating in a plentiful supply of fresh grape juice makes a perfect day for them. They begin to reproduce slowly at first and then at increasing rates until they reach an exponential peak with that great display of bubbling that overflowed the crock. Saccharo E are eating sugar and eliminating ethyl alcohol and carbon dioxide gas with great abandon. They see no end to their good fortune; they are Masters of the Universe.

But soon they have consumed more than half the supply of sugar and as suddenly as they increased, they begin to decline. Their environment is changing rapidly; natural selection spreads out its impartial arms and individual by individual, the less efficient stop reproducing, starve, and die. Their dead bodies slowly settle to the bottom of the jug. Not only is the supply of sugar decreasing, but also the concentration of alcohol is increasing. Most Saccharo E are killed by a solution containing more than 16% alcohol and 13% is often the limit. They die starving and smothered in their own waste. With the end of bubbling, their colony becomes extinct.

As each batch of wine is completed, Saccharo E. yeasts seem to become extinct in their universe. That would be true for a closed universe but a winemaking event is not closed. During the explosion of carbon dioxide gas that so impressed us, many Saccharo E individuals escape in the bubbles. They spread into the larger universe. Many find other sources of sugar and start the process over again. All those who remain behind soon die. Except for this early escape from the process, the species Saccharo E would have become extinct long ago.

Hopefully, the experiences of Saccharo E can provide some insight for us. No living creature can sustain unlimited increases in population indefinitely. Darwin's observation that living creatures produce more offspring "than can possibly survive" is a characteristic of life that humans must adapt to or eliminate if they are to survive in the long term. However, there are some signs that humans will be able to control their population growth. The Chinese have chosen the one-child family as the best model for their society; some Western European countries are experiencing population decreases. A few industrialized countries with well educated populations show slower growth. These are all hopeful trends. However, we must keep our wits about us; no species has been able to sustain increasing population growth for long.

The Limits of Economics

Capitalism is one of the economic models for producing and distributing goods and services necessary for the maintenance of human life. Be assured that the previous sentence is about as accurate and complete a statement as you will ever find in the immense literature on business, capitalism, and economics. Nowhere will you find a straight forward list of the Principles of Capitalism. Instead there is only an endless discussion of what some facet of economic activity does or is thought to do (Heilbroner and Thurow, 1975, 1984).

But a short list of topics that continually appear would include: supply and demand; scarcity, goods and services; ownership, production and distribution; profit and loss; and division of labor; also tradition, command, market prices, commodities, labor, wages, trends, land, rent, and capital investment by individuals or government.

A few comments might explain how these concepts are used in economic affairs. The purpose of any economic activity is to produce or find and distribute the goods and services necessary for human life and procreation: food, clothing, shelter, safety etc. Goods and services are related in a supply/demand ratio; as supply goes up, the demand generally goes down. The more dresses or pants you have, the less you need another. For long stretches of time, 80% of humans lived and labored on the land. Humans have always traded goods and services depending on how each perceived the relative value and need of various possessions; this is the "invisible hand of the market" adjusting the price or cost. Labor can increase its productivity by specializing in specific skills. Ownership of land, labor, and capital can be used to invest in organized efforts to produce profits.

The three economists and their masterworks in the following list, developed the underlying theory of capitalism, particularly the *laissez-faire* form of capitalism as practiced in England around 1800 and barely changed in the USA of today.

1776. Adam Smith — *Wealth of Nations*

1798. Thomas Malthus — *Essay on the Principle of Population*

1817. David Ricardo — *Principles of Political Economy*

Capitalism, as described and defined by Smith, Malthus, and Ricardo was a collection of inherent self-regulating variables resulting in an equilibrium that encompassed all members of society. Adam Smith called this mechanism "the invisible hand of the market place."

The England of Adam Smith's day has been described as a "dog's hole" (Heilbroner, 1962). Extreme poverty and wealth, child labor, workhouses, and high infant mortality rates led to that conclusion. But it was all justified by the effortless operation of the Invisible Hand that automatically provided goods and services. Everyone was driven by narrow self-interest, which ensured that he or she attained a proper position in the social scale. If a product became scarce, competition would direct more resources to its production. If there was a shortage of labor, wages would go up and people would have more children, or encourage immigration. When specialization was shown to produce more output per person or if costs went up, investors and owners would buy equipment to increase production and lower cost while increasing profit per piece; profits would be accumulated to buy more equipment, etc. Everything operated on the law of supply and demand by an unrelenting and invisible market mechanism based solely on self interest.

But long before economists such as Adam Smith provided us with a theory, the earliest humans based on the family unit during the era known as the hunter/gather period provided the goods and services needed within the technological abilities of their time. The male killed animals with spear, arrow, and club; the female gathered roots, seeds, and berries while performing her major function of bearing and caring for the young. The early family unit already contained the division of Labor, which Adam Smith was to use much later to explain the rapid advances of living standards in England. The hunter/gather period has been estimated as beginning more than 1,000,000 years ago.

The humanoid population was very small during this period. When the food resources of an area declined, the hunter/gathers would simply move on to new hunting grounds. Originally there was no need to own territory but as the population increased and extended families began to split up, conflicts arose over possession of hunting grounds with a resulting increase in male brute force. This led to the early militarism of tribes and clans. As human imagination and skills developed, new information and techniques became available helping primitive people to gain more from their environment.

As the population grew clans, tribes, and societies emerged. Control and ownership of land by the militarily powerful became the norm.

Human Inventiveness

Ever since Adam Smith and his magical Invisible Hand, economists have been touting their economic principles as the cause of improving standards of living. That is a complete absurdity and false; the source of higher standards of living is the discovery and accumulation of observable technical and scientific knowledge. Economic activity creates little more than the organization and focus of workers on processes and technique. Human activity and creativity provide the basis for the production and distribution of goods and services for the maintenance and improvement of life and society.

Human inventiveness appeared with the first human creatures over the last few million years. Hominid tool-making abilities are well known. Below are a few dates documenting the sequence and cumulative impact of inventions long before the Industrial Revolution began in 1750 and before Adam Smith's *Wealth of Nations* was published in 1776. Economists describe how an economic system functions and how to profit from it; they do not create an economic system that necessarily benefits everyone. Early dates are approximate and are often set back in time as archeological findings grow (Grun, 1982; Ochoa & Corey 1995).

BC Time Period

2 MYA	*Homo habilis*, "handy man," stone scrapers
1.8 MYA	*Homo erectus*, exodus from Africa, spears
230,000 BC	Human control of fire for heat and cooking
120,000 BC	*Homo sapiens*, stone tools and points
30,000 BC	Bow and Arrow, cave paintings
13,000 BC	Humans learn to make fire (previously they only "found" it)
10,000 BC	Herding begins, with goats in what is now Iran
8,000 BC	Agriculture invented, in what is now Iraq
7,000 BC	Clay pottery, mortar in Jericho, woven cloth in what is now Turkey
5,000 BC	Copper metal knives, irrigation
3500 BC	Sundials, plow, cuneiform writing in Sumer
3300 BC	Wheel invented in Sumer
2900 BC	Stone building, huge pyramids, Egypt
2800 BC	Bronze armor and spears
2300 BC	Weights and measures, shekel, cubit, foot, Sumer

2000 BC	Chariot invented, wooden ships in Crete
1400 BC	Iron production, by Hittites in what is now Turkey
1100 BC	Use of coal as fuel, China
750 BC	Architectural arch in use, hand cranks, by Etruscans
450 BC	Idea that every event has a natural cause, by Leucippus
100 BC	Glass blowing in what is now Syria

AD Time Period

50	Rotation of crops
100	Magnetic compass needle, China
270	Gun powder, sulfur and saltpeter, China
537	Dome design, Hagia Sophia church, Constantinople
700	Porcelain invented, China
1098	Chinese water clock
1282	London cathedral clock, Exeter 1284, Oxford 1288
1396	Gothenburg printing press
1492	Large sailing vessels. America discovered by Europeans.
1500	Waterwheel by Leonardo Di Vinci
1589	Knitting machine invented
1596	Thermometer by Galileo
1600	Treatise on Magnetism by Gilbert
1609	Systematic movement of planets explained by Kepler
1615	Coal used for energy as wood becomes scarce in England
1620	Experimental scientific method formalized by F. Bacon
1639	Micrometer
1643	Barometer by Torricelli
1652	Air Pump
1657	Pendulum clock by Huygens
1660	Microscope invented (Leeuwenhoek)
1662	Boyle's gas law — $PV = nRT$
1662	Royal Society of London (King Charles II)
1666	Académie des Sciences, Paris (Colbert)
1668	Reflecting Telescope (Newton)
1675	Velocity of Light (Romer)
1690	Steam piston pump

1712 Newcomen steam engine

1733 Shuttle loom

1765 Watt steam engine

1771 Oxygen discovered (J. Priestly)

If we add up the progression of events listed above, the creativity and increasing scientific and technical inventiveness of humans becomes obvious. This is a very abbreviated list of advances because, of course, print space, too, is a limited resource.

Modern science really began about 500 years ago with:

1543 Copernicus, N., *Revolutions of the Heavenly Bodies*

1618 Kepler, J., *Epitome Astronomiae*

1620 Bacon, F., *Novum Organum*

1638 Galileo, G., *Discourses Concerning Two New Sciences*

1687 Newton, I., *Philosophiae Naturalis*

These life changing scientific works were written well before the Industrial Revolution (1750–1900) and Adam Smith's *Wealth of Nations* in 1776. These sequences of dates prove beyond doubt that growing human capabilities in science and technology predated, caused, and fueled the Industrial Revolution. Should these advances begin to falter or slow down, the increase in living standards will also slow down, particularly if population continues to increase.

Limits of Capitalistic Economics

1. Capitalism adjusts to changes in supply or demand by moving prices up or down quickly. Limits are essentially based on human needs and desires which are often daily or immediate.

2. When limits occur, capitalism searches for alternate materials and distribution to fill the need.

3. Capitalism has an inherent tendency to monopolize any limited supply of resources using patents, wealth, specialized equipment, secrecy, mergers, acquisitions, trade barriers, military power, inertia, etc.

4. It is limited to advances in science and technology for growth.

5. It cannot produce more after consuming the limit. It cannot create more land, water, oil, coal, uranium, energy.

6. It cannot repeal diminishing marginal return limits.

7. It is tolerant of extremes of poverty or wealth.

8. It has no inherent ethical principles. Its motivating force is profit without limit.

9. It is limited to the creative abilities of its workers. Its most important function is organizing human activities.

The claim, wherever it is made, that the improvement in human standards of living is due to economics, not due to the advances of science and technology is a falsehood. Economic Capitalism, by itself, has little to do with uncovering new scientific facts unless there is the prospect of immediate large profits; then it might foster the application of technology.

Many of our firmest beliefs are based on absolute concepts. The theory that the invisible hand" of the free market will correct every limit to supply and demand is ambiguous as well as false. In the sense that prices will adjust to supply, it is meaningful, but to imply that the invisible hand will correct a limit in the supply of oil or anything else is false. As we are now witnessing, as oil becomes scarcer, the invisible hand of the market will only increase the price until people use less of it, or technology finds more of it, or science finds a substitute. The implied meaning of the invisible hand concept is that it will correct the problem. It does no such thing. When the price goes high enough, consumers will be able to use none of it. The invisible hand by itself creates nothing; it never has. Inherent in Adam Smith's Hand argument is the inference and assumption that there is a limitless supply of all commodities, from labor and land to water, copper, gold, oil, gas, coal, uranium, and tar sands etc. These are completely false assumptions.

The economists of the Simon School (Simon, 1981, 84, 96) attempt to refute the limits defined above. These economists declare unequivocally that there are no limits in resources anywhere to be found on Earth. Humans will find more oil, land, edible plants, and anything else they need or want. Such false opinions are the result of looking at numbers that relate to human concepts, not at observable natural phenomena.

Limits of Science

Science is limited to finding and describing what is in nature to be found and described. That is an ultimate limit. It is limited to those areas of natural phenomena, which can be repeatedly observed by some means. Scientists are always searching for more data and better experiments and these take time. The 109 atomic elements took several centuries to discover, for instance. The progress of science is inherently slow because it is dealing with what is unknown, ambiguous, and sometimes beyond imagination.

From hydrogen through uranium there are 92 natural atomic elements listed by chemists. There is another rare or unstable group of about 17 that have been briefly created or identified in various laboratory experiments (Chang, 1988). Will we find another 92 abundant and stable elements in the coming years? No; that is not a possibility because of the weakening of the electron-proton orbital attraction with increasing separation. In fact, the most widely distributed elements are the first forty lighter elements in the periodic chart. Who needs another 92 anyway? The answer is: economic progress needs it. Every limit to scientific discovery, limits the ability of capitalism to exploit new technologies that could lead to the introduction of new products for profit.

But there is always the possibility that someone might find unique materials that vastly increase the chances for innovation. Buckyballs, which are groups of 60 carbon atoms bonded together in the shape of a ball, were discovered in 1995. But no one seems to have found an important use for them as yet. Recently (May 30, 2006 issue of *Proceedings of National Academy of Sciences*), Dr. Lai Sheng Wang, a physicist at Washington State University, announced the discovery of gold in the shape of hollow cages of 16 gold atoms and 13 gold atoms in the form of sheets. The 16-atom cages have triangular faceted exteriors. Dr Wang hopes to put other elements into the interior of the cages, perhaps creating new materials with uniquely valuable characteristics.

Suppose one could take a lump of iron and cheaply change it into rust-free threads or tubes that could then be used to reduce the weight of iron structures by one half; would that be significant? Absolutely; the limits of science and technology would be permanently expanded and innovation would have a new lease on life. But the above scenario is not a likely one and would certainly have limits of its own. Scientists have been investigating and describing every nook and cranny of nature with progressively decreasing findings of dramatic material significance.

It is noteworthy that around 1900 it was said the limits of science were known and that science's remaining task was to measure everything out to the third or fourth decimal place. We are still not there yet. The discovery of the structure and information content of DNA is one instance of how far we have come since then and how far we have yet to go. But after all the caveats, the general proposition still holds; we have entered an era of diminishing marginal returns from science and technology that directly impacts standards of living. That is not to say the advances of science and technology will come to a halt, they will certainly continue to diffuse and expand into the third world. Newer developments such as the understanding and manipulation of DNA for human benefit are still critical areas for research and development as are the vast reaches of the Cosmos.

Science attempts to describe the natural universe; it can say little about the subjective worlds. They cannot be definitively described or understood scientifically because they occur only once or erratically and often have only one observer.

The methods of science can be easily perverted or subverted by human error or motives for fame, wealth, or plain ignorance. Recall the Piltdown hoax. Recently there was a report from Shanghai (*New York Times*, May 15, 2006) that a US-trained Chinese scientist at Jiaotong University had stolen a computer chip design from a foreign source and passed it off as his own. Chinese society was understandably shaken by the news but scientists often have selfish motives, like other humans.

Societies That Survived

Although Diamond described many societies that collapsed, scattered throughout the book are references to five societies that survived for relatively long periods of time essentially intact. These include Tikopia, Tonga, Highland New Guinea, Iceland, and Japan before 1868.

Tikopia is an island in the Southwest Pacific Ocean which has a population of about 1,200 people that lived on only 1.8 square miles of arable land for about 3,000 years (Firth, 1957). The Tikopia purposely limit their population by abortion, infanticide, suicide, celibacy, and *coitus interruptus*. Because of high infant mortality rates, a primitive society's survival often depends on large families; however, under conditions of finite land and food resources, an increase in the number of large families soon surpasses the limits of sustenance. As one of the few known people that controlled population growth so completely, the history of the Tikopia clearly shows that a society can avoid collapse from excessive population increase.

Another survivor is Tonga, a Pacific archipelago with a large population that existed for over 3200 years on 288 square miles of land. Diamond cites Tonga as a top-down approach to survival. It has a centralized government with a king who protects the interests of the whole island group since he owns it all. Therefore, he is alert to any practice, such as deforestation that causes erosion and takes action to correct the problem.

A third example of a society that was able to survive is in the highlands of New Guinea, which is the large island north of Australia. In the 1930s a plane flew over these highlands and to the amazement of the crew and the world, the area was covered with farms, which were likened to the farm areas of Holland. When overland expeditions reached the highlands and explored the area they found people had lived there sustainably, for about 46,000 years when New Guinea and Australia were united as one land mass. The people inhabiting the area had migrated eastward from Asia by way of Indonesian islands.

The first traces of agriculture appeared about 7,000 years ago, along with deforestation up to 8,000 feet and artificial drains in valley marsh. By way of explaining the drains, note that New Guinea is tropical and has up to 400 inches of rain per year. Overall, Europeans found that the highland people were "primitive"; they used sticks for digging, felled trees with stone axes, and lacked metal. But they were extremely sophisticated in their farming techniques.

Using a process of bottom-up experimental adaptation to their site, the New Guineans sustainably exploited their environment. After initially deforesting the area, they began to reforest by transplanting fast-growing native ironwood seedlings, *Casuarina oligodon*, to their gardens to provide fire wood, construction timber, nitrogen fixation by the roots, and leaf fall for compost. That is in addition to the fact that trees help to prevent landslides on steep highland slopes. In effect, the highlanders invented what we now call Silviculture, the growing of trees instead of field crops for human purposes.

The highlanders do not have strong political institutions; the basic unit is the small local village; the only communication is by foot and word of mouth. But they suffer from the same Darwinian–Malthusian problem all species face. That is: "As many more individuals of each species are born than can possibly survive," taken from the Introduction to *The Origin of Species*. Somehow population must be controlled; did highlanders control their population? Yes, they used the usual procedures of war, infanticide, herbs for contraception and abortion, and abstinence. They thereby avoided the fate of the Easter Islanders.

Iceland is the forth area Diamond cites as a survival society. It is a large island off the coast of Greenland near the Arctic Circle. When the first Viking settlers arrived from Scandinavia and Britain they had no inkling of the fragile nature of the soil compared to their native lands. Using their homeland practices, they soon deforested Iceland.

The Vikings were noted in the medieval world as the scourge of civilization; using their light and fast boats driven by sail and oar they appeared suddenly, apparently out of nowhere. Often arriving as traders, they soon discovered that sword, shield, and robbery were much less time consuming and more profitable than trading.

The Vikings started out from Scandinavia into the North Atlantic along a string of nearby island groups, the Orkney, Shetland, and Faeroe Islands. The Orkneys were conquered about AD 800. The Shetlands and Faeroes were settled later in the 800s. All three island groups formed societies that have been sustainable for the last 1200 years.

Iceland is much larger and farther west in the north Atlantic near Greenland. It has many volcanoes and hot springs that Icelanders now use to heat their houses. Although farther north than the island groups and Norway's main farmlands, because of the nearby Gulf Stream, the climate is moderate. Initially, one quarter of the island was forested but the early settlers of 870–930 quickly cleared the forests by 80% for timber and pasture. Today, Iceland is about 1% forested.

Iceland has very fragile soils that form slowly and erode easily compared to the Viking's familiar robust soils in Norway. The settlers did not understand that fragility until too late. Now Iceland is the most ecologically damaged country of Europe. The population grew quickly but there were many local famines because of the large variations of weather in the late middle ages. However, about that time a rise in dried cod fish trade began that rejuvenated the society until today Iceland is one of the richest in the world on a per capita basis.

The fifth survival society Diamond cites is Japan before 1868. The year 1868 was the year of the Meiji Restoration which began the conversion of Japan from an isolated self-sufficient kingdom into a modern industrial state. It also restored the Emperor once again to the revered symbol of the national character that He had been in ancient times.

Japan is a crescent shaped string of islands curving offshore in the Pacific Ocean from along the most easterly finger of Russia south toward Korea. The Japanese language is polysyllabic, akin to Korean and languages of the Altaic region of northwest Asia, not monosyllabic like Chinese. The early tribes migrated from the Korean Peninsula to their present island homes after originating in central Asia.

Eighty percent of Japan is sparsely populated, forested mountains; the remaining twenty percent is coastal plain where food production is concentrated. Birth rates rose and fell with similar moves in rice prices in early Japan because of food shortages. Families attempted to restrict births by marrying later and nursing longer, as well as contraception, abortion, and infanticide.

The southern four islands form a group around an inland sea where the civilization began (Fairbank et al., 1973 and Packard, J. 1987). The first historic emperor was enthroned around AD 400. World War II Emperor Hirohito was reputed to be the 124th inheritor of the Imperial title Tenno, literally Son of Heaven.

A second state power emerged in 1603 when the Emperor, remaining as a figurehead in the old capital of Kyoto, made Tokugawa the chief of state, or Shogun. In 1615, the Tokugawa clan took full control of Japan operating from its base in the city of Edo (modern Tokyo). As the dominant military–administrative power the Tokugawa controlled the country through 259 local military officials, the Daimyo.

From 1650 to 1853, Japan was a closed society; only the Dutch were accepted as carefully controlled traders. With peace and seclusion, population soared, doubling in a century. By 1720 Tokyo became the world's largest city.

The five societies sketched above show clearly that human populations can exist for long periods of time if growth rates are restricted. Societies that did not limit population, such as the Easter Islanders, Mayans, and Harappans, collapsed leaving hardly a trace. We are now in the required limited population phase. Hopefully, we can establish a balance within the limitations of Earth and avoid a similar collapse.

The Diffusion of Science

A cultural phenomenon that has been evolving for more than a century has largely escaped general notice. Modern science, which had its source in Western European society and was central to Western European ascendancy over all other nations, is now rapidly diffusing throughout the third world. The United States participated in Western science with such early scientists as Edward Morley, Willard Gibbs, and Asa Gray. But Japan was the first non-Western nation to understand and implement Western science and technology. The Meiji Restoration (1868) imposed modernization on Japanese society that transformed its educational, scientific, industrial, legal, and military cultures (Kennedy, 1987). The Restoration began partly as a reaction to a perceived threat fifteen years earlier by Commodore Perry's United States fleet of black steel ships arriving in a Japanese port (1853). After only a few decades, Japan demonstrated its newfound technical abilities by destroying the Russian Far Eastern fleet at the battle of Tsushima near Port Arthur in 1905. That is only 37 years for a nation to absorb and use the advanced findings of science and technology.

Full demonstration of Japanese mastery of science and technology came at Pearl Harbor in 1941 and in the postwar period as Japan became a major competitor to all nations in semiconductors, autos, and industrial technology, quality, and efficiency.

Now, after the challenges of Japan, the Asian Tigers, and Russia, we are faced with the real behemoths: China — four times our population, and India — three and climbing. We might think this is threatening. But these nations are not striving to take anything away from us; they want to have what we have, an improving standard of living based on the application of science and technology. This trend will determine the tenor of the times for the indefinite future.

Bibliography

Alberts, B. et al 1998, *Essential Cell Biology*

Bacon, F. 1620, *Novum Organum*

Chang, R. 1988, *Chemistry*

Copernicus, Nicholas (1473–1543) 1543, *The Revolutions of the Heavenly Bodies*

Grants, B.R. & P.R. 1989, *Evolutionary Dynamics of a Natural Population*

Deedy, J. 1972, *The Complete Ecology Fact Book*

Dennett, D. C. 1995, *Darwin's Dangerous Idea*

Diamond, J. 2005, *Collapse*

Fairbank, Reischauer, and Craig 1973, *East Asia*

Firth, R. 1957, *We, The Tikopia*

Galileo, Galilee 1638, *Discourses on Two New Sciences*

Gould, S. J. 1989, *Wonderful Life*

Grun, B. 1982, *The Timetables of History*

Heilbroner, R.L. 1953, *Worldly Philosophers*

Heilbroner, & Thurow 1975, 1984, *The Economic Problem*

Heinberg, R. 2005, 2nd. Ed, *The Party's Over*

Hubbert, M. K. 1969, *Resources and Man*

Kennedy, J. 1987, *Rise and Fall of the Great Powers*

Kepler, Johannes (1571–1630) 1618, *Epitome Astronomiae*

Keynes, J. M. 1936, *General Theory of Employment*

Malthus, T. 1798, *Essay on the Principle of Population*

Marx, K. 1867, *Das Kapital*

Mayr, E. 1991, *One Long Argument*

Meadows, D. et al., 1972, *Limits to Growth*

——, 1992, *Beyond the Limits*

——, 2004, *Limits to Growth: 30-Year Update*

Newton, Isaac (1642–1727) 1687, *Philosophiae Naturalis*

New York Times, 5/15/06 "Computer Scientist's Fall"

Ochoa, G. & Corey, M. 1995, *The Timeline Book of Science*

Ricardo, D. 1817, *Principles of Political Economy*

Schopf, J.W. 1999, *Cradle of Life*

Simon, J. L. 1981, 1996, *Ultimate Resource*

Simon, J. L. & Khan, H. eds. 1984, *Resourceful Earth*

Smith, Adam 1776, *Wealth of Nations*

Strickberger, M. W. 2000, *Evolution*

Tainter, J. A. 1988, *Collapse of Complex Societies*

Veblen, T. 1899, *Theory of the Leisure Class*

Chapter 3. Life is Optimism Triumphant

Heedless of hazard or fear of failure, life presses on with indomitable ardor. Life is optimism triumphant. It is the perfect foil against an unknown future. Carbon-based life at its very core is naked optimism created by little more than star bursts and star light, atoms and photons. Life will live and reproduce, no matter what the obstacle. It knows only the incredible optimism of being, regardless of time, place, or change.

This is the time to leave behind the subjective delusions and mythical universes with their dark dread of death to take a courageous place in the natural universe of observable fact. These illuminate the path to the truer and an understanding of life. For too long we have wallowed in absolutes, confusion, and ignorance while stumbling along a path leading toward disaster and extinction. We, who now are the natural selectors of life on Earth, celebrate the unique, awe filled, and enchanting optimism of being.

The Natural

The natural accepts humans as part of observable nature. The natural is open to all in the vast and growing stores of biological fact and theory in chemistry, physics, astronomy, geology and on and on. Science is free for all to confirm or deny. Natural science and its handmaiden technology, for better or worse, have become increasingly the basis of human ascendancy over the biological and physical environments of Earth. Our lives are now totally dependent on understanding the findings of the sciences and applying their technologies to solve the present and rapidly emerging problems of nature and humanity.

Our Companions

Our companions of the past gave us love and life. They struggled to live and formed a bridge of being, the gift of life, to us. We hope to do the same. We must honor them all; if not, we are diminished.

Our yearning for quick and easy solutions to our emerging problems of limits and global warming belittles our ancestors. They were given no purpose, no meaning. We must find the way as they did along the hazardous passage from bacteria to man. We alone can determine the purpose and meaning of life because we are Life, uniquely conscious Life that is also optimism incarnate. Species have no life, struggle, pain or death. At extinction they simply end or fade away. But we individuals of mass, energy, space, and time; live, endure, and finally die with optimism and hope, alone. Yet the individual being is the key and cornerstone, the "atom" of life. Without individuals there would be no creation of life, no species, no evolution, and finally no humans.

The Methods of Science

The first traces of science appeared with the ancient Greeks, Chinese, and others sometime before 500 BC. But modern science really began with Copernicus (1473–1543), Galileo (1564–1642), Kepler (1571–1630) and Newton (1643–1727), less than five hundred years ago.

The startling images these men uncovered are contained in their masterworks:

- 1543 Copernicus, N., Revolutions of the Heavenly Bodies
- 1618 Kepler, J., Epitome Astronomiae
- 1638 Galileo, G., Discourses Concerning Two New Sciences
- 1687 Newton, I., Philosophiae Naturalis

Modern scientists in their efforts to understand and explain natural phenomena use observation, imagination, induction, intuition, dreams, hunch, deduction, scientific fantasy, inferences, and anything else to find a good idea or hypothesis. Once conceived, the hypothesis is put through all possible experimental and observational tests to determine agreement with established science and then to determine whether new facts can be deduced from the hypothesis and confirmed. Almost all hypotheses fail but if there is one that withstands all comparisons and tests, it is considered a new theory that can be absorbed into the overall body of science as long as no new data contradict it. Although deduced facts certainly lend credence to a hypothesis, alone they cannot yield proof because there is always the possibility of better information surfacing. Complete acceptance of new facts and theories must be delayed until additional scientific laboratories with competent scientists duplicate and confirm the original investigation.

New theories are held rather lightly until confirming data overwhelm all doubt. The scientific method lacks the deductive shield of non-contradiction to protect and prove it, but it is by far the most fruitful and efficient method of logical analysis and confirmation of fact uncovered to date. Unfortunately, science progresses slowly and scientists are often like boxers fighting an invisible and devious foe. A bare description never tells of the fog they fight through at the frontiers of the unknown. But the results often seem obvious after they are uncovered, explained, and published.

A comparison of possible truths to find the better one is the core of the scientific method. When a scientific theory is proved, it is truer than any other theory examined. There is no claim of absolute truth or the truest, only the truer among the possibilities under consideration at the time. Everyone is invited to produce facts that prove or disprove a theory. This is the unique strength of the scientific method; it is open and self-correcting. There are no philosophic absolutes that endure for long. The tests of comparison, contradiction, and confirmation give scientific theories their proof. Under the umbrella of such relatively modest claims, the panorama of experimental-theoretical science has grown and blossomed for the illumination and benefit of all humanity. Science is owned by no one, rather it is free to all who will learn and use it.

The scientific method as now developed combines sensory observations and reasoned hypotheses followed by confirming or falsifying testing to reach tentative conclusions — human truth. Compared to Plato's deduced Absolute Essences or Ideal Truths, which fostered an enduring confusion, the scientific method achieved for mankind an unsurpassed level of knowledge that has led directly to immense benefits for human health, longevity, and living standards in an amazingly short time. Unfortunately, when scientific technology is combined with laissez faire capitalism the result is many of the growing problems and crises that face us today. These can be solved only by more advanced scientific research and technology. For example, consider the universal problems associated with global warming, increasing pollution, erosion, and a population explosion leading to a climatic consumption of finite limited resources. Can they be solved by wishful thinking, denial, and spin?

Scientific explanation is the second way our bodies know the truth. The first is the knowledge embedded in the helixes of our DNA. This is inherent unlimited optimism, with knowledge far beyond our own. It can duplicate itself and its host.

DNA brings every protein or enzyme it encodes to each cell or molecule whenever needed for maintenance or growth. It will take centuries if not forever for us to learn as much. Living cells must maintain themselves and reproduce, otherwise life ends. DNA optimistically assumes continuation without qualification; there is nothing to fear, nothing to anticipate. Each cell presses on regardless of anything or nothing. "I have everything; I need nothing; I am invincible." Living beings accept life with no end. Only we conscious creatures can see an end to life, but we seldom take notice for optimism suffuses us, too.

Evolution and Natural Selection

There is general confusion as to the relationship between the theories of evolution and natural selection. Darwin's original formulation of evolution was a combination of at least five sub theories (Mayr, 1991). These included evolution in general, descent from a common ancestor, multiplication of species, natural selection, and gradualism. Evolution is the umbrella theory, and natural selection one of its parts. Today there are a number of additional theories supporting evolution. These include gene mutation, genetic recombination, reproductive isolation, the molecular structure of DNA, and the information content of DNA. All are scientifically consistent with both the Evolution and Natural Selection theories.

Darwinian evolution is the theory that all living species descended with modifications from earlier species going back to the origins of life on earth about 3.8 billion years ago. Confirming proof of this is the relatively recent discovery that all life forms, from unicellular bacteria to multi-cellular plants, fungi, and animals use the same molecular DNA chemical composition and structure and the same general processes of reproduction. The theory of evolution draws on information from all scientific fields for confirmation. It cannot be disproved without disproving other independent sciences.

Natural selection is difficult to confirm experimentally because of the long stretches of time involved compared to a human lifespan. However, some bacteria and molds reproduce in days or hours and these have been observed to change into new species when grown in environments of increasing stress (Strickberger, 2000; Alberts et al, 1998).

Natural selection is a single, but paramount, part of overall evolutionary theory. Darwin concisely explained it in the introduction to *The Origin of Species* in 1859:

> As many more individuals of each species are born than can possibly survive; and as, consequently, there is a frequently recurring struggle for existence, it follows that any being, if it vary however slightly in any manner

> profitable to itself, under the complex and sometimes varying conditions of life, will have a better chance of surviving, and thus be Naturally Selected. From the strong principle of inheritance, any selected variety will tend to propagate its new and modified form.

"The survival of the fittest" and "the struggle for existence," which are often used as shorthand for natural selection, are not accurate descriptions of the process. Darwin described these phrases as metaphorical only. They are often interpreted to mean that only the fittest survive the struggle. This is a false interpretation. In a benign environment all or most individuals survive (Dennett, 1995).

The most spectacular example of survival within a benign, open environment is the remarkable expansion of the human population over the last few hundred years as they occupied new territory in the Americas and Pacific and learned to control the dangers of harsh environments by improving food production and health procedures. The human population has increased from about 100 million after the last ice age ten thousand years ago, to one billion in 1804, to about 6.5 billion now (Noble, 1972). The only significant drops in the European rate of population growth occurred after the collapse of the Roman Empire c. AD 476 and during the Black Death of 1350.

Natural selection selects nothing. It is simply a description of the events that can occur when a creature's replicating genetic structure varies and is incompatible with an unpredictably varying environment at some arbitrary moment in time. The environment gives no choice or exception or explanation to anyone. It is simply there; it actively seeks and selects nothing. It has no end purpose; it never has.

In fact, some of the events integral to natural selection have little or no violence associated with them. The bisexual evolutionary process requires that individuals both maintain life and reproduce. Half of this requirement, the need to reproduce, is seldom violent. The bonds between male and female, parent and child, and individual and clan are all nonviolent and advantaged by empathy, love, and cooperation.

The Gifts of Natural Selection

From a human point of view, life is the primary gift. The heart and energy of life is optimism. Life's creatures never have the time to know or doubt the future; to stay alive they must hurry on. Without knowledge of the past or future, they press on into the present for whatever it contains. Enthusiastically life embraces each moment to add to its collection. It is inherently certain that something of value is out there somewhere. But that which is of paramount value we have already, the incredible optimism of being itself.

The early earth had a chemically reducing atmosphere that was mainly water, hydrogen, sulfur oxides, hydrogen sulfide, and carbon oxides. Early life forms found ways to use these reducing compounds as energy sources for their metabolic needs. About 3.5 billion years ago (BYA) photosynthetic bacteria evolved that could convert water, sunlight, and carbon dioxide into glucose (carbohydrates) and free oxygen. The evolution of life that could use free oxygen metabolically resulted in an increase in available energy that made more complex organisms possible. Photosynthetic bacteria increased the oxygen content of the atmosphere from less than one percent to twenty one percent. They are still, together with green plants, actively supporting us with oxygen. For more than three billion years they have been producing this truly miraculous gift; without it, plants, animals, and humans would not exist.

The creation of an oxygen rich atmosphere was the first instance of living organisms causing overall major changes in the physical conditions on Earth. This took place mainly from 3.6 to 1.8 BYA. The second instance is the present era beginning about 200,000 years ago with the evolution of modern humans. Can anyone doubt that we are making major changes in the conditions for life on Earth now?

Other primary gifts of natural selection are the evolution of bisexuality and with humans, consciousness began.

Bisexuality and Ethics

A major change in evolutionary direction came about with the evolution of bisexuality about 1.5 BYA (billion years ago) then flowered in the Cambrian explosion of 500–570 MYA (million years ago). Bisexuality led directly to the loving parents, particularly the loving mother who carries and nurtures the young, protects, and sustains them. The intense emotional feelings of altruism, empathy, love, and compassion began here. That is not to say human families began here; 1.5 BYA there were no mammals. But ancient starfish for instance, which lacked even a backbone, draped their bodies over their young for protection until they were self-sufficient (Wendt, 1965).

Unbidden, our ancestors replicated themselves and created us. From within the confines of their DNA instructions they never selected, they won or lost with weapons they had never used, against foes never seen. And the independently varying environment might at any moment hurl a single high-energy photon that disrupts the molecular structure of a gene causing death, disease, or mutilation. But regardless of endless dangers, within the genes of all individuals there stands a defiant shield of optimism, firm against all odds — "we will continue undeterred and unafraid for we will live."

Darwin in the Descent of Man (1871, 2nd. ed. 1874) noted that "any animal whatever, endowed with well-marked, social instincts ... would inevitably acquire a moral sense ... as soon as its intellectual powers had become as well ... developed as in man." Considering scientific information now coming into print, Darwin's analysis of 1871 was remarkably prescient.

Rational moral behavior can favor the preservation of life forms and thus be "selected" by natural selection. From the family and kin, love and altruism spread to flock, herd, clan, and society. The prairie dog's alarm call to alert his companions of a circling hawk sometimes attracts attention to the alarm caller with fatal results. That is certainly an altruistic act. In general, altruistic behavior by individuals can lead to improved survivability of the species and thereby to a permanent place in the DNA of individuals.

The single-cell division form of reproduction (mitosis) continued exclusively until about 1.5 billion years ago or for about the first 2 billion years of life on Earth. Then somewhere and somehow an individual or several individuals divided into two slightly different beings, each unable to reproduce until they combined their DNA with another — the defining sexual act. Until this place in evolutionary time, individuals were independent and complete. They lived and reproduced by dividing into two identical cells, alone. Bisexual species had a strange new burden to carry. They consisted of two kinds of members — male and female. They could not reproduce alone; they were incomplete. Nevertheless, this was the evolutionary path new creatures like plants and animals would follow. Here, at this juncture, are the beginnings of love, family, clan, and the wellspring of ethical behavior.

Life had set out on a new path. For a long time there was little notice in the fossil record of any great change. Then seemingly in a brief period about 500–570 million years ago, bisexual, multi-cellular life became abundant during the so called Cambrian Explosion. New kingdoms of life — plants, animals, and fungi appeared for the first time (Gould, 1989).

A Rational Life

A rational life must have as a basis the evolution of life on Earth. We evolved from subconscious mammals. They are unaware of their existence and behave according to instincts embedded in DNA. Innumerable individuals died while capturing and embedding this knowledge in DNA. Now, as the only self-aware conscious creature, we can reexamine these instinctive behavioral characteristics. Are such DNA instructions compatible with the present needs of humanity, life and Earth? DNA is not capable of adapting quickly enough to save us from the dangerous limits that loom. Instead we must put our trust in rational behavior that will allow us to pass safely through the present hazards by controlling some of our animal nature, particularly greedy and violent behavior embedded in DNA. This we must do to create a human society that brings us into a self-sustaining harmonious equilibrium with each other and all of nature. We must learn to follow new cultural imperatives that include all species and our common habitat Earth.

As the only conscious, rational creature on Earth, we must awaken to our responsibility to follow a code of behavior that transcends exclusively human interests. An objective rational life, designed to foster the continued existence of life and humanity should be derived from and focused on the realm of objective natural reality not on Plato's subjective absolutes and essences.

Life Is Optimism Triumphant

As the only conscious creature on Earth we recognize death while standing at the very pinnacle of life. Our question becomes; will we be worthy of those ancient creatures who prepared the way than gave us life? It is also Hamlet's well known question (act lll, scene 1):

> "To be, or not to be — that is the question..."

Every creature that ever lived, without pause and unthinkingly gives the answer to Hamlet's question: "I will live to the limits and beyond regardless, for I am optimism triumphant."

The triumph of optimism indeed! Leibniz, the eminent philosopher argued that this world is the best of all possible worlds. Voltaire's Candide popularized that view by mocking it. But evidently Leibniz never walked in the forest. If he did, he didn't notice that the ground was littered with dead leaves, branches, and tree trunks. Yet those green Oaks still standing in beauteous splendor carry on as if endowed with eternity. And thus all forms of life press on no matter what; they continue with inexorable ardor to the last fleeting moment of optimism.

Stand in Awe

Consider the way things are.
As the keystone of the arches
We stand at the center
Taking the measure of all things
Absent, the universe is some other.
We take the detritus of nova and make of these:
The will to live,
The gift of love,
The search for truth,
Reverence for life,
Compassion toward all,
The judge of the universe,
With Cosmic vistas
And changing events
As the flea worries the dog,
So worry we the Cosmos
For against all odds,
We are here.
Never meant to be
We are yet something more than nothing,
Optimism triumphant!

Bibliography

Alberts, B. et al 1998, *Essential Cell Biology*

Ayers, A. J. 1990 *The Meaning of Life — Essays*

Babbitt, I. 1936, *The Dhammapada*

Cupitt, D. 2003, *Life, Life*

Darwin, C. 1871, 2nd Ed, *Descent of Man*

Davies, P. 1999, *The Fifth Miracle*

Grants, B.R. & P.R. 1989, *Evolutionary Dynamics of a Natural Population*

Dennett, D. C. 1995, *Darwin's Dangerous Idea*

Diamond, J. 2005, *Collapse*

Gould, S. J. 1989, *Wonderful Life*

Katz, L. D. ed. 2000, *The Evolutionary Origins of Morality*

Mayr, E. 1991, *One Long Argument*

Meadows, D., et al 1972, *Limits to Growth*

Noble, P., Deedy, J. 1972, *The Complete Ecology Fact Book*

Schopf, J.W. 1999, *Cradle of Life*

Strickberger, M. W. 2000, *Evolution*

Wendt, H. 1965, *The Sex Life of the Animals*

Part II. Aspects of Physical Reality

Chapter 4. The Search for Truth

The character and uses of the truth are obvious; we use them every day. But do we really know the truth that represents and conforms to the real world? When we see a red barn and say to a friend, "there is a red barn," are we telling the truth? Our friend understands our meaning and agrees — "there is a red barn." But the barn will have nothing to do with the color red; it rejects red; it reflects red. Red is not a part of the barn. The barn absorbs all other colors and incorporates their energy into increased atomic vibrations that warm the barn roof and siding, but red it refracts and reflects away. We should really say, "There is a barn that rejects red." And so it is with all other objects we see — the yellow peony, the red cardinal, the green oak. Similarly when we look at a star or galaxy, we do not really see an object; we see the light that it emitted perhaps five or ten billion years ago. The star or galaxy could have ceased to exist by the time we "see" it. When the ancient architects of our language described a dark cave as black and a leaf as green they considered color to be an inherent part of the object. With our present scientific understanding of light and color as secondary, independent aspects of objects, we have a different view. Our ideas of truth change as science unfolds and human knowledge increases.

Our bright stars and red barns are part of the common sense view of the world or naive realism as the philosophers call it. But common sense does not satisfy those who have studied the character of truth most carefully. They point out that our awareness of truth is dependent on sensory organs and reasoning brains and that these have significant limitations. After all, on earth, distances are so short compared to the velocity of light (186,000 miles per second or 3 x 10^{10} cm. per second) we seem to see earth objects instantaneously. Within the far reaches of the Cosmos the distance light travels in a second is utterly trivial. We now know that starlight often must travel billions of years to reach us here on earth. We can never know the position of a star simultaneously with the position of our earth; we both moved during the flight of the photon. We must consider these objections and raise some questions of our own. What can we know? What does it mean to know? How do we know? What is it we really know?

Beginning in a conscious way with the ancient Greeks about 3000 years ago, we can trace increasing interest and effort in the search for truth. However, at each stand along the way, we insisted and knew absolutely that we had The Truth. Since we are always certain The Truth must be protected and propagated, endless atrocities and wars resulted. This was the outcome of The Truths of nationality, race, and economics. In such fields intolerance is endemic and violent even when clothed in ethical and rational motives. These intangible and abstract Truths hold us with the firmest grip and imbed in us the virus of fanaticism. It is the character of absolute truths, whatever dogma they wear for camouflage, to bring us into evil.

We have only to remember the cyclic expansions and contractions of state borders, tribal savagery, wars, intolerance, and economic exploitation to feel the hypnotic power of The Truth. And yet over time, many of these Truths yielded to truer truths. We can expect the same for our Truths. Here we use the capitalized word Truth to stand for the absolute truth that in words and reason corresponds perfectly with reality and has no exceptions. These absolutes are the source of intolerance in every guise. Yet, if history is a guide, we knew only the truer truth at the time, nothing more and often less.

In science, which in its modern form has existed for only a few hundred years, many True ideas have already been abandoned including phlogiston, ether, Lamarckism, Ptolemy's earth-centered universe, Aristotle's basic elements — air, earth, fire, and water, blending inheritance, continuous radiation, Darwin's gemmules and so on. This short list could be extended indefinitely but scientists have a clever habit; once they notice a phenomenon well enough to name it, new insights are absorbed under the old name. Recall the two ancient Greeks Leucippus and Democritus who decided a rock could not be broken into smaller and smaller pieces indefinitely; a limit must eventually be reached. That limit was the Greek *atomo*, now known as our atom. Slowly over 2500 years the idea of the atom has been by turn novel, opaque, mystic, limited, denied, expanded, and then generally accepted in the Western world around AD 1900. The atom is still evolving (Pullman, 1998), but it is not the smallest thing anymore. Now we have protons, electrons, quarks, neutrinos, mesons and so forth, all much smaller than the atom.

The Limits of Human Truth

It is one of the great tragedies of civilization that from ancient Ur, Anyang, Thebes, and Harappa to Modern Westernized societies, none have found a sure way to determine the truth. None have stood the test of time and change. We have found no logical, sensory, or intuitive method to find the Truth or even lesser truths with certainty. There is no certainty — things change. Philosophers, prophets, priests, and professors, among the most talented and dedicated of humans, have spent thousands of lifetimes in the search with little success and nature herself is unyielding. Our brains and five senses, after slowly evolving on our three dimensional planet for almost four billion years, are still limited in scope and defective in application. Yet as the only tools we had in our long struggle for life, love, and offspring, they served us well enough. We are here.

Human Sensory Limits

Our senses are limited. Most humans hear frequencies as high as 20,000 cycles per second. Dogs, bats, and dolphins do better and we now have electronic detectors that do better still. The ear's sensitivity to frequencies within the normal range varies a great deal and as we age our ability to hear high frequencies declines rapidly.

The situation with regard to vision is much worse. Sight depends on visible photons (electromagnetic waves) reflected, refracted, or radiated from the object we are viewing. Electromagnetic wave phenomena cover an immense frequency range up to 10^{20} cycles per second. Human eyes are sensitive to a minuscule portion of this total spectrum, from $4.3–4.7 \times 10^{14}$ cycles per second. We cannot see or detect by any direct means electromagnetic waves such as radio, microwave, infrared, x-rays and gamma rays. For these we use modern sensors and counters such as Geiger counters, gauges, antenna, and amplifiers. Our vision is also subject to the deficiencies of farsightedness and astigmatism. If we press a finger against the side of an eyeball, we see double; if we take certain drugs, we can also see double or more. If we put a stick into water, it bends where it enters the water. On deserts we sometimes see an oasis where there is none.

Well-known scientific laws can now explain all of these deceptions. For instance, the bent stick is not bent; it is the refracted and reflected light waves from the immersed portion that have been bent. Snell's law of 1621 exactly explains this. Light is bent (or refracted) when it travels from a medium of low density (air) into one of higher density (water) or the reverse. Similarly, the mirage is due to light bending as some waves move through layers of dense cold air and some through less dense warm air. Why do our eyes not compensate for refracted light? Probably because a bent stick or a desert oasis are seldom life threatening; therefore, there is little survival value involved.

We seldom realize when we think of truth, facts, knowledge, or understanding, what a huge number of things we are considering. We don't normally think of truth as having a quantitative component. But if we could take the combined knowledge of the professors at a large university, for instance, and put it all into one brain, it would undoubtedly have thinking capabilities far beyond any other. Yet such a brain would have only a tiny fraction of the factual content of the university's library. Brains have serious limitations as to how much information they can detect and store and how much they can retrieve and compare at any one moment. Recall your school play; wasn't it hard to remember your lines?

Then there are other limitations. When we learn a new truth, do our brains automatically adjust all our stored truths like a computer spread sheet can adjust related numbers? Also there is an input problem. How can we get all the worthwhile information into our brains? In answer to that question, we spend ten or twenty years at school but still have mammoth deficiencies. In addition, we are time limited; the average person reads much less than one book a month and squeezes in a look at a newspaper, magazine or television on the fly. Then too we seem to forget facts and truths almost as fast as we accumulate them. Also we seem to prefer some truths more than others and remember them better regardless of their truth content. These are probably the source and containers of our individualized belief systems.

When we add up all these limitations it becomes obvious that we can never know the absolute truth about any significant problem. Absolute, in this case, means that beyond which there is nothing; nothing exists; there are no qualities, no quantities, and no changes. Of course, some will say that the laws of logic, 1=one, for instance, are absolutely true; and so they are, but these are definitions, and definitions are only repetitions using other words. They add nothing new to our knowledge. What our brains really know are the observed human truths that have their basis in the truer of possible truths at the moment, these we shall call truer truths.

The Search for Truth

Although China and India had developed forms of logic beginning about the 5th century BC, ancient Greece is the source of Western civilization's logic traditions. Aristotle (about 350 BC) and his school were foremost in the field, particularly in deductive logic and the syllogism. Inductive logic was relegated to a minor role and considered of dubious importance. Much later Francis Bacon, for instance, took exception to some of Aristotle's methods in his *Novum Organum* of 1620. It took us from 350 BC to AD 1620 to begin to realize empirical data and detailed observation are important. Theoretical analysis based on deductive logic alone could get us only so far; observation and systematic experimentation are essential to understand the nature of reality.

Because of the magnitude of the subject, we will be able to cover only a sliver of the thought and literature that has gone into the search for truth, but hopefully we can come to understand some of the problems involved.

The Deductive Method, The Syllogism

Although the ancient Greek philosophers used reason of all kinds in their everyday activities, they were mesmerized by deductive logic in their writings. Formal deductive logic, correctly applied, can yield Truth. The strategy is to start with accepted Truths (called premises) and determine from them the Truth of the conclusion. Unfortunately, the concluding Truth is usually part of what we already know or is implied by the premises. If we take apart a deductive form, the syllogism for instance, we can see how the strategy works (E.B. 23:251, 1989):

Every dog is a mammal.	—First premise.
Some quadrupeds are dogs.	—Second premise.
Therefore, some quadrupeds are mammals.	—Conclusion.

The general form of this syllogism is:

Every X is a Y.

Some Z are X.

Therefore, some Z are Y.

Whenever arguments or statements are put into this form, the logical procedure is correct and the conclusion is valid for the particular premises used. We have arrived at the truth. But is it necessarily true? It is true only if the first two statements, the premises, are true. How can we know that the premises are true? To prove each premise we need two more true premises and so regress endlessly.

Can we use inductive methods, which do not give absolute truth? Must we examine every X in the universe to determine if they are all Y and also if some Z are X? This is well beyond our capabilities. But dogs, mammals, and quadrupeds are not wandering around the universe; they are confined to the earth. They are limited and defined by humans to stand for specific classes of X and Y. The truth of the above syllogism depends on the definitions of relationships among the animals. The class of mammals, by definition, includes quadrupeds and the class of quadrupeds includes dogs (but not birds, for instance). We are merely reciting the man-made hierarchy of definitions of classes that we already know or can find in a dictionary or text book. We have found no new truth. We learn from the syllogism only what is explicitly contained in or can be inferred from the unproven premises; there is no new knowledge created in the conclusion. The philosophers get around this inconvenience of unproven premises by stating the premises are self-evident truths.

In addition to the syllogism there are also three laws of logic and a number of rules of inference that have played important roles in the development of deductive thinking (Hospers, 1965). The capital letters P, A, and Q are often used to represent propositions, arguments, and statements. The three laws are:

The law of Identity — if P is true, then P is true.

The law of Non-contradiction — not both P is true and P is false.

The law of the Excluded Middle — P is either true or false.

We agree with these three laws because all three are built into our language and reasoning abilities. However, the law of the excluded middle is suspect. If John says he is at position A and not at position B, we can look and see that he is at A, not at B and is therefore telling the truth. In modern science there are many cases where P is not true nor false — only possible or probable. For instance, if P stands for the position of an electron in an atom, the excluded middle law does not hold. An electron in an atom can have one of many possible positions. We cannot say at any particular moment that an electron is at position A not position B. There is a possibility that an electron can be at A or at B or somewhere else. The likelihood of an electron's position at any moment is describable by a probability function. Each possible position can be given a percentage probability that an electron occupies that position. In addition the very concept of position is a bit fuzzy according to Heisenberg's uncertainty principle (Heisenberg, 1958). The principle states that some pairs of properties, such as position and momentum or energy and time, cannot both be precisely determined at the same time. This is an intrinsic property of the quantum world not a human measurement limitation. The probability function is an integral part of Quantum Mechanics theory, which is one of the most powerful and successful scientific theories of all time.

If you insist that the electron is either at position P or it is not at position P, you can only get the reply (from our present knowledge) that the position of an electron obeys a probability function. The electron will probably be at P, zero, or ten, or thirty or some other percentage of the time. This concept of the probability function has been successful in predicting the behavior of particles, waves, atoms and chemical reactions. Perhaps we should add to the law of the excluded middle the statement: P must be either true or false, or probably true or false.

The Nature and Limits of Inference

To infer is to derive by logical reasoning an additional conclusion from the meaning of an original statement or observation. It is a form of deduction that can be used to enhance a statement. For example: "the boy runs down the hill," what can we infer from that? One, the land behind is higher than that in front of the boy. Second, the boy is healthy; he can run. Third, he is running for some reason. Perhaps a bear is chasing him; perhaps he runs to meet someone; perhaps he is training for a place on the school track team; or he is overjoyed about some good news. We can't be certain. The point is we can infer conclusions by reasoning about a statement regardless of its specific content. But the inferences are not necessarily true; they can even contradict each other. Inferences are not proofs.

Scientists use inference in every step of the scientific method. In the inductive phase when looking for a good idea or hypothesis they try to infer evidence from subjects related to the area of investigation. After forming a good hypothesis they test it by observation and experiment and then deduce or infer new concepts to test and compare to established science. But the inference must be experimentally tested before it can be considered as scientific. Some of the rules of inference are (Hospers, J. 1965):

If P implies Q, and P is true, then Q is true.

If P implies Q, and Q is false, then P is false.

If either P or Q is true, and P is false, Then Q is true.

The word implies in these rules is equivalent to the word infer. The problem with these inferences is the word if; we must know beforehand that P is true before we can be certain of the truth of the conclusion. Only empirical observation of the factual nature of P can give us that information. So we are left with uncertainty again.

The rules are ambiguous and misleading. In the first rule, the word if before the second P is missing. The important qualifier should be first and; therefore, the rule should read:

If P is true, and P implies Q, then Q is implied true.

The other rules should be put into the same form. Again the premise has not been proven; therefore, the conclusion is not proven.

Another danger in the use of inferences is that we often infer inference from inference until we have a chain leading to nowhere or to falsehood and have lost sight of the original premise.

We must accent the distinction between an empirical observation of a fact and an inference from the observation. An observation is the use of our five natural senses to create a perception in our brains. A perception is a mental construction, an idea, a model of reality. It is not the empirically observed fact. An inference is a subjective human mental construction, called appropriately a construct. It is not a fact, although we often treat it as one. Note that an inference can be drawn from any form of reasoning including: induction, deduction, probability, imagination, lies, or fantasy.

If we use these constraints properly, we can, for example, observe that all things are attached to the earth somehow; we can then (with the help of Isaac Newton) infer from a mental construct or model that there is a force we call gravity holding us on. And so it is with all other mental constructs; they are not facts only subjective feelings, intuitions, or beliefs. Of course these can be very important to us but they are not facts and we should not claim them to be or use them as facts. Again inferences by themselves are not proofs (Graziano & Raulin, 1989; E.B. 6: 308, 2a; Hospers, 1965).

The denial of a law of logic or rule of inference would lead to self-contradiction. This is the ultimate test that authenticates deductive laws and rules. But is self-contradiction alone enough to prove or disprove a statement? There is a certain statement by a Cretan, Epimenides, that "Cretans always lie." If the statement is true, then as a Cretan, he lied by stating the truth. If the statement is false then the Cretan lied originally.

Formal deductive logic statements are self-consistent and play a dominant role in such fields as geometry, arithmetic and computer electronics. Who can deny that 1 = 1; that is indeed The Truth. The AND, OR, NOT, NAND, and NOR logic circuits which in various combinations constitute computer logic are essential to the functioning of all computers. Deductive logic is required whenever we reason from general statements to specific ones. We could say the general idea in some way contains the specific idea; we are uncovering something that is already there. But in long, involved arguments, deduction often helps to keep our thinking clear. In science, deduction also plays an important role, as it does in our everyday reasoning. Nevertheless, deduction alone cannot guarantee truth unless the premises are true and of that we have no assurance.

The Axiomatic Method

In addition to the syllogism, the laws of logic, and rules of inference, the ancient Greeks also gave us the axiomatic method of obtaining the truth. An axiom is a statement or relationship that is self-evident and accepted without proof much as the two premises of a syllogism are accepted without proof. Euclid's compilation of the *Elements* of geometry from 300 BC and earlier was the example that established the axiomatic method. Through more than two thousand years of history the method was accepted as the perfect model of a deductive path to true theorems.

Euclid proposed ten self-evident axioms and from these proved four hundred and sixty five theorems. Theorems are statements that are thought to require proof, unlike axioms. In most editions of the Elements, only five of the axioms are related to geometric proofs:

1. Only a single parallel line can be drawn through a point outside a given line.
2. A straight line is the shortest line that can be drawn between two points.
3. A straight line can be extended indefinitely.
4. All right angles are equal.
5. A circle can be constructed given its center and a point on it.

In the nineteenth century, doubt was thrown on the parallel line axiom when it was shown that self-consistent geometric formulations can be laid down by assuming that more than one parallel line can be drawn to a given line through a given point or that no parallel can be drawn to a given line. The first applied to straight lines drawn on curved surfaces, as on a sphere, and the second with a saddle shaped surface. From these developments it was realized that a valid non-Euclidean geometry could be formed from any set of axioms that does not yield contradictory theorems (Paulos, 1991).

Into these growing doubts about the true role of mathematics, stepped an Austrian mathematician, Kurt Gödel. In 1931, at age 25, he published a short paper that is arguably the most important achievement in logical thought of the twentieth century (Paulos, 1991; Hofstadter, 1979; Nagel & Newman, 1958).

Gödel showed in his first incompleteness theorem that a formal system of mathematics is always incomplete. There will always be pertinent statements that cannot be proved or disproved. Later, in his second incompleteness theorem Gödel proved that no set of axioms could demonstrate its own consistency. We can only assume the internal consistency of such a set.

The consequences of Godel's theorems for mathematics and logic are enormous and still under analysis. Initially, some thought the theorems apply only to human language and brains but the theorems allow no distinction between human thought and thinking machines; they both must follow the same principles. We are left with a deductive logic, which has no more validity than inductive logic-they are both tentative and dependent on the truth of their premises or the extent of their facts. This does not mean logic is worthless; it means we must carefully assess the truth of every statement and belief. The things we know are often limited in strange ways but particularly by the finality of absolutism.

The Inductive Method, Mendel's Peas

Inductive logic can guide us from specific, observed events to a generalized model of all similar events. The specific events must be examined repeatedly so the observer can study their characteristics in detail. Induction; however, does not yield absolute truth because it is not self-evident in the deductive sense and usually cannot be shown to cover all possible events. For example, Mendel's study of how adult pea plants transfer their traits to young plants was based on thousands but not every pea plant in the universe. First he studied peas for two years before choosing seven different traits for experimental testing. The selected traits included "pure" strains of tall versus short plants, white versus red flowers, and smooth versus wrinkled seeds. The "pure" strains of peas were then crossbred and their young crossed again. After the first cross, only one of the traits appeared in the offspring from each pair of traits — it was called the dominant trait. The missing trait was called recessive; where did it go? At each step thousands of plants were examined to determine the number of young with the original traits. After the second cross, the dominant trait was exhibited by three of the young plants for every one showing the recessive trait, which had reappeared. Where had it come from? In the recessive state the trait is overshadowed and hidden by the dominant trait. Mendel thought he could understand more about pea plants if he observed many of them. He was correct even though he had not studied every pea plant in the world; his findings were remarkably successful in extending our understanding of the inheritance of traits in plants and animals by showing that every trait of a bisexual creature (eukaryote) is controlled by two possible factors that are in pairs in the chromosomes of the double helix. The traits are physically based on molecular arrangements called genes that are parts of the chromosomes. However, in the XY chromosome, which determines the sex of a creature, the shape and chemistry of the two chromosomes are markedly different.

Reasoning from the particular to the general is an inductive process; from the abstract and general to the more specific is a deductive one. The inductive thinker's task is to examine pertinent bits of information while trying to imagine a process, inference, or model that explains them all. He or she cannot examine every possible bit of information in the world because there are too many; therefore, the conclusions can never be totally proven. If a good idea or hypothesis is found, it is then compared to all other established concepts. Agreement with related concepts leads to the acceptance of the idea as probably true.

The Scientific Method, The Source of Knowledge

The scientific method takes an additional step beyond inductive logic to support the truth or falsity of the hypothesis. This step is to deduce or infer new facts from the inductive hypothesis that possibly could be observed experimentally. If new deductions are found and confirmed, the hypothesis is considered true until a better one surfaces. It is generally thought that the scientific method is an inductive method that can explain large amounts of data. It is that but also contains deductive parts. Scientists use imagination, induction, intuition, dreams, hunch, deduction, inferences, and anything else to form a "good idea" or hypothesis. Once conceived, the idea is put through a series of observational tests to determine agreement with established science and then to determine whether new facts can be deduced from it and confirmed. If it passes all tests, the hypothesis is considered a new theory that can be absorbed into the overall body of science as long as no new data arise to refute it. The scientific method might better be called the inductive-deductive-test method.

The scientific method tries to turn deductive logic on its head. At the conclusion, if deduced facts are confirmed, they are used to prove and authenticate the original hypothesis. Although deduced facts certainly lend credence to the hypothesis, they cannot yield absolute proof because there is always the possibility of uncovering new and better information. Conversely, in the syllogism form of deduction, we begin with the limitations of self-evident but unproved premises and use them to prove the final conclusion.

Scientific theories are accepted as long as no better theory or contradicting information comes along. New theories are usually held rather lightly until confirming data from other laboratories overwhelm any doubt. The scientific method lacks the deductive shield of non-contradiction to protect and prove it but it is by far the most successful and productive method of logical thinking we have ever found.

A comparison of possible truths to find the truer truth is the basic process of the scientific method. When a scientific theory is proved we are really saying the theory is truer than any other theory we have examined. There is no claim of absolute truth, only the truer truth among the possibilities under consideration. This is the unique strength of the scientific method; it is open to self-correction. Everyone is invited to produce facts that disprove the theory. There are no ideological absolutes that must be met and endure indefinitely. The tests of contradiction, contrasts, constraints, and confirmation give scientific theories their proof. Under the umbrella of such relatively modest claims, the panorama of experimental-theoretical science has grown and blossomed for the illumination and benefit of all humanity.

The Expanding Universe

It sometimes happens that we cannot determine which of two inductive hypotheses is truer. In that case, we must wait for more confirming or conflicting experimental data to become available.

During the nineteen fifties and sixties there was a growing controversy about the origin and character of the universe. Although many were involved, two prominent astronomer cosmologists were at the center of the controversy, George Gamow (1952) and Fred Hoyle (1950). In simple terms, Gamow favored creation of the universe from a great expansion of energy and mass beginning from a single point of extreme density. Gamow's insight was that since Einstein's relativity equations were able to accommodate Hubbell's expanding universe, they could also apply to a contracting universe. Going back in time, this would finally lead to a universe condensed into a point of extreme or infinite density — a singularity. On the other hand, Hoyle advanced the theory that the universe was continually producing matter by condensation of an undetected, dilute background material throughout interstellar space. There was no definitive experimental evidence that favored one theory over the other. But looking back, we can see there were a number of indications in the 1930s and 1940s that would have thrown considerable light on the debate (Gribbin, 1996).

By 1960 there were several international groups looking for the leftover radiation from the great expansion predicted by Gamow. Oddly, two scientists at the Bell Laboratories in NJ accidentally found it in 1963. They were working on a huge microwave horn antenna for use in satellite communication and had a problem with interfering radiation, which was observed to be white noise or static coming from every direction in space. Strenuous efforts to clean and adjust the antenna were fruitless. Finally the two scientists, Arno Penzias and Robert Wilson, consulted with Robert Dicke and his colleagues at Princeton University, who happened to be working on an apparatus to search for this type of radiation. The noise Wilson and Penzias had found was apparently the background radiation predicted by Gamow and others as a result of the initial great expansion. It had a temperature or energy of about 2.7 degrees Kelvin, as predicted by Gamow's theory. Here was the confirming experimental information that doomed Hoyle's continuous creation theory. But he left a lasting imprint on the successful theory anyway; Hoyle was the one who first called it "the Big Bang."

The best description of the source of the 2.7 degree radiation begins with the early Big Bang. Initially the universe consisted of extremely high-energy radiation. With continued expansion simple particles such as electrons with a few protons and neutrons condensed from the fiery cauldron of energy. The photons were of extremely high frequency and energy, with short wavelengths. As the universe continued to expand it cooled, but the photons were constantly scattered, absorbed and reemitted internally by the particles. Later, when the temperature had dropped enough, electrons, protons and neutrons began to coalesce to form the simple atoms hydrogen and helium. The photons were now relatively free from scattering and absorption by particles. With continued expansion, the wavelength of the photons increased. They can be thought of as s t r e t c h i n g — as they lengthened in expanding space and shifted toward longer red wavelengths of decreased energy content. After traveling for billions of years as the universe expanded, in 1963 some of them entered a microwave antenna in New Jersey manned by Penzias and Wilson. They had the relatively low energy of 2.7 degrees Kelvin as predicted. The big bang theory had experimental confirmation; the continuous creation theory had none. Note that the important word here is experimental. Many assume that the great theories are the major achievement of science. Certainly theories are of great import, but experimental facts are the foundation of science not theory. Theories often give us wondrous overarching understandings that encompass huge amounts of data. But it is the workaday accumulation of data that play the deciding role in the selection of one theory over another. Theories are tentative human constructions; data are forever.

The Probable Truth

The ancients were fixated on THE TRUTH — the Absolute Truth. But when man began to study random and seemingly chance events, a new kind of truth began to emerge — the probable truth. When an event has several different possible outcomes, we can assign to each outcome a number representing its probability in each particular measurement. Probability studies have been growing in all fields of human endeavor including: quantum and statistical mechanics, quality control, chaos theory, gambling, insurance, social activities, and measurements on evolving systems of all kinds. Mendel's work on the frequency of traits in pea plants, described elsewhere, is an early example of the use of probability theory in the field of biology.

The classic example of probability is the flip of a coin. It will land either heads or tails; therefore, each possibility is fifty percent. In probability theory, zero is taken as zero percent probability and one as one hundred percent probability. The coin flip is then ½ for heads and ½ for tails. This does not mean heads and tails will alternate. Each flip is independent of all others so there can be strings of heads or tails. But over a large number of trials, the head will appear 50% and the tail 50% of the time.

The Trial and Error Method

The trial and error method is the ancient, common sense predecessor of the scientific method. It is nature's method and has been used since time immemorial in such processes as natural selection and DNA evolution. Trial and error involves comparing different outcomes and then accepting the most useful at the moment. All life depends on a correct response to trial and error situations. This is the pragmatic approach. When a caveman saw a lion growing larger and larger, from instinct and tribal experience he turned and ran. When a frog sees a black spot on a floating leaf, he flips out his tongue to grab a fly or, if the spot grows larger, he dives.

When we turn on the television and nothing happens, the first thing we do is bang the side of the cabinet — the amateur repairman's faithful solution. If that fails, we check the plug to the electric outlet; if it is all right, we connect a lamp; if it does not light we check the circuit breakers. If they are all right, we call an electrician. This is the trial and error method we use throughout the day, every day. It is plain common sense based on our experience and reason. It is also the inductive–deductive, scientific method in a rather casual disguise. In this way, we are all scientists and always have been.

Do animals use the trial and error method? When a herd of elk grazes, they follow fresh grass wherever it leads. They learn about the color green. They find that the forests or high mountains have little grass, particularly in the winter, and avoid them. The human herder similarly takes his sheep to fresh pasture in a regular seasonal pattern he or his ancestors learned through trial and error. We often attribute such behavior to instinct, but it is like the instinct we use when we go to the butcher shop rather than the green grocery stand to buy meat. We rob banks rather than homeless shelters because, as a successful robber once told us, "That is where the money is."

The evolutionary process of natural selection, discussed elsewhere, is also a trial and error method. Briefly, natural selection is contingent on two unrelated factors, variation and selection. The trial part comes about when, for example, the Galapagos Island finches vary in their ability to crack hard seeds. In an increasingly dry, rainless environment only cactus, which typically has thick shells, can produce seed and only the individual finches with strong beaks can crack them. The weak billed and the strong billed are on trial for their lives. The strong billed live; the others were born with the wrong bill; they were in error for that dry environment and starve to death (Grant, 1989). This seems inhuman, but the natural selection process is unconcerned with individual death and provides no corrective solutions in terms of individual lives. Only species live on for any long period of time.

The evolution of DNA (deoxyribonucleic acid) from a small relatively primitive molecule to the enormously complex molecule that controls the development of every living creature today is also a product of trial and error. Because of chance mutations or copying errors, the molecular structure of DNA varies somewhat. Individuals with such variations must stand trial before the filter of natural selection. If the molecular variation is helpful, the individual will survive and prosper, if deleterious, the individual could die or leave no offspring. It is an error to be born with the wrong genes.

The Truer Truth

We return, finally, to the questions asked at the beginning: what is truth; what can we know; how do we know; what does it mean to know; since we can believe the false as well as the truth, what is it we really know?

What can we know? We can know what we directly observe and consider by reasoning. Our brains have memory and reasoning abilities; our senses bring signals into our brains for comparison and logical reasoning. We make a point-by-point comparison between closely related possibilities to determine which is truer; we form a mental construct, a model. Here is the basis of the truer truth. We learn by comparisons, by finding differences and contradictions. When we consider two possible truths, we compare them part by part; the one that is most compatible with and least contradictory to our memories and observations is truer; it is our truth for the moment. That is all we know, all we can expect to know and, on balance, all we need to know. It was the truer truth that guided us and our companions through the hazards of our evolutionary journey. As a corollary, human truths change with time.

When we look for an answer to a problem whether philosopher or average person, we are looking for the truest true solution. We consider all the possibilities we can find at the time and compare each with the others. If we learn that some are false, we reject them. One by one we discard the less true until there is one or none left. If none, we hunt for other approaches to the problem, if one, we accept it as truer. For the moment it is the truest truth. This is the thinking process we use every moment, every day whether we are searching for a new suit or a new method to make semiconductors.

A truer truth is simply the result of a choice between comparisons of information collected from the senses and the memory. Comparisons and choices are made by reasoning until only one truth is left standing — that is the better truth, the truer truth.

What is truth? As with many things — atoms, water, universes — truth is a human creation couched in human terms to help us survive and understand our lives. Wherever we look we see truth, but at any moment we see only a little bit and a little way and what we see is often changing as we look.

Our judicial system is based on finding the truth. Yet after the professionals have completed their composite of deduction, induction, authority, gossip, hyperbole, and hunch, an amateur jury of citizens is given the task of determining the truth of the charge. Why is it that a group of average citizens is given the task of uncovering the truth? The jury's method is to find by comparison, which set of facts has the truer truth. Sometimes mistakes are made. But it is the best we can do at the moment and seems to have a relatively high probability of accuracy. Thus at the highest and cruelest level of societal action — crime and punishment — we are reduced to the truer, relativistic solution to finding the truth. We recognize that absolute truth — The Truth — is unattainable unless we restrict our belief system to the deductive method of reasoning. It is certainly The Truth that if A equals B and B equals C, then C equals A. The deductive method is of some value in determining the consequences of two possibly true premises but it leads to no new truth of its own and no help with the unknown.

Truer truths can arise by comparing old observations with new. Vesto Slipher observed in 1912 that the hydrogen light spectrum from the Andromeda galaxy was bluer than the same atomic spectrum from nearby stars in the constellation Hercules or here on earth. Later, in 1929, Edwin Hubble showed that the more distant a galaxy, the more red-shifted its spectrum. Here was proof that the universe was expanding. In a static universe there would be neither red nor blue shifts. By noting the difference in spectral wavelength images on two photographic plates, two sets of data were obtained from the past and present that could be compared in Doppler's equation to determine the velocity and direction of stars and galaxies. Since Slipher's Andromeda spectrum was bluer, it was approaching our Earth.

Truer truth can be decisive in any human event. In war the victor is often the side that knows the enemy's intent and plan more exactly. When the Japanese Admiral Nagumo launched an air attack against the island of Midway on June 4, 1942, he assumed the American carriers were still in Hawaii. That was part of the plan handed down by the Japanese General Staff. But the Americans had broken the Japanese naval code, knew exactly where and when the attack would come, and had their carriers waiting a few hundred miles northeast of Midway. By the time Nagumo knew three American carriers were nearby, 150 planes were already on their way to attack him. After almost total destruction of the attacking American torpedo aircraft, the dive-bombers sank three Japanese carriers within five minutes and the remaining one shortly after. The fleet carriers Naga, Horyu, Akagi, and Soryu sunk that day were four of the six that had attacked Pearl Harbor on December 7, 1941. The Midway action partially redressed the Sunday morning attack on Pearl Harbor, but more importantly, it was the turning point in the Pacific war.

The Japanese had decided the Americans would stay in Hawaii until they received reports that Midway was under attack. Reconnaissance efforts to confirm the whereabouts of the American carriers prior to the attack were complete failures. The searches included long-range air sorties over Hawaii, a cordon of submarines between Hawaii and Midway, and Nagumo's aerial search to the east immediately before the attack. The long-range search was canceled for lack of refueling capability; the submarines arrived at their positions two days too late, and some of Nagumo's search planes had mechanical problems (Fuchida and Okumiya 1955). Despite the lack of critical information about the whereabouts of American forces, the Japanese pressed forward with invincible ardor. Their thought was that by the time the American fleet reached Midway, Japanese troops would have occupied it and Nagumo's fleet would be waiting to destroy them with superior numbers.

A firm opinion as to how your enemy will act begs for disaster. Events do not always happen the way we decide they must. The Japanese plan was based on superior numbers, skill, and conjecture. The American plan was better; it was truer.

Although the Japanese had more aircraft, ships, and experience, the Americans had one advantage: they had broken the Japanese code and had truer information about the coming battle.

An analysis of two specific events or statements of truth is simply a search for the statement that has a greater content of tested knowledge. We evaluate truths by their differences and freedom from contradictions with what we already know. In this way we compare things from the price of tomatoes at two different venders to speculative variations between human opinions.

What does it mean to *know*? On our human subjective level, truth is a feeling of familiarity, consistency, confirmation, acceptance, or satisfaction for one possible truth from among the alternates under consideration. It is without conflict, confusion or the frustration of contradiction. But a light does not go on or a gong sound when we know something; all we have is a feeling of harmony. We judge the meaning of truths by comparison of differences and a choice for the better. There is no assurance of truer truth before we learn the consequences of our choice. The constant testing of truer truth is one of its basic advantages. When the test is successful, we store the information for future use, when wrong we begin anew. We live on a diet of assumption and hunch with qualifying trials in between.

To find the truer truth it is necessary to have creditable, alternate ideas to compare. Here, education and continued study show their importance. The more knowledge, the more possible ideas we have from which to choose. Ignorance can be a total obstacle to the emergence of the truer truth.

The choices we make often pit what we know from memory against new facts we have found. This creates a conflict. What we know from memory, we hold with confidence because of earlier tests and agreements. We have affection for these well confirmed memories and favor them against anything new. That is the natural conservative tendency of all living creatures. Also we have the ability to hold thoughts, beliefs and attitudes in separate compartments that have not undergone the process of comparison. By such methods we are able to believe almost anything, regardless of contradictions or conflict. Facts in memory often contradict better replacements, yet they can be held with the tightest commitment and emotion. We can also forget that a new fact might be in conflict with other peripheral beliefs that also need adjustment. If the new fact is a major one, our total mental equilibrium can be disrupted and we can sink into confusion or apathy. For these reasons it is often more comfortable to stay with beliefs in a Pascal-like wager; better to believe a profitable falsehood than an apparently profitless truth.

We know what our five senses and reasoning powers seem to give us, but what is it we really know? Remember the red barn that had little to do with red and rejected it? Similarly, what we know on the easily perceived naive realistic level is often necessary for our survival but is only a first level of reality. It is the reality in which we evolved and in which our five senses and brains evolved. Below and above this there are other levels. Some of these are the scientific realities uncovered over the past few centuries and which have led to our dominance as a species. There is also the emotional level — fear, greed, love and hate; the aesthetical level; the religious level; the communal level; and so forth. We often hold the inferred consequences of these levels of thought in total isolation from one another.

Authoritarian and Absolute Truth

How can we live and function with only the truer truth to guide us? Must we believe in absolute truths to have meaningful and truthful lives? Remember, absolute truth was defined earlier as that beyond which there is nothing observable. Our answer is that life continues and has always lived with only the truer truths that allowed it to adapt to the many variations that confront it. But beyond these truer truths humans have often believed in absolute truths that automatically reject conflicting facts. With such mixtures of truth, truer and absolute, we live in constant conflict and bewilderment. Being human we often fearfully choose the Absolutes when we stand alone before a magnificent and incomprehensible universe that finally takes our lives away. We accept truer truth when we must live and function in a changing world. They cannot both be true; we must make a choice or live in continued conflict and error.

Between the concepts of truth and truer truth there is a vast chasm. The Authoritarian or Absolute Truth is without exception and eternal. Truer truth is of a human evolutionary scale, tentative, and often restricted to a small subject and a short time. A major difference is that truth does not evolve, the truer truth does. In conclusion we must understand that whatever its nature, truth is not "out there" for us to absorb whole in a moment. Human truth of any kind is an internal mental transformation and model based on electrical signals from our senses to our brains, not external reality. Our brains and senses have limited capabilities and were not made to know, nor needed to know, ultimate reality. Each human holds somewhat different truths. Again we must draw the distinction between believing and knowing. We can believe almost anything witness, extraterritorial visitors to Roslyn, predestination, spirits, devils, samsara, witches, perpetual motion machines, and endless other unobserved or unobservable events. We know only what we can observe or accept from a carefully evaluated observer. We can and often do believe anything that might serve our inclination.

Bibliography

Encyclopedia Britannica, 1989, designated volumes

Fuchida, Mitsuo and Okumiya, Masatake, 1955, *Midway, The Battle that Doomed Japan*

Gamow, G., 1952, *The Creation of the Universe*

Grant, B. R. and P. R., 1989, *Evolutionary Dynamics of a Natural Population*

Graziano, A. & Raulin, M., 1989, *Research Methods — A Process of Inquiry*

Gribbin, J., 1996, *Companion to the Cosmos*

Heisenberg, W., 1958, *Physics and Philosophy*

Hofstadter, D., 1979, *Gödel, Escher, Bach*

Jospers, John, 1965, *An Introduction to Philosophic Analysis*

Hoyle, F., 1950, *The Nature of the Universe*

Hume, R.E., 1959, *The World's Living Religions,*

Leplin, L. ed., 1984, *Scientific Realism*

Nagel, E. and Newman, J. R., 1958, *Godel's Proof*

Paulos, J. A., 1991, *Beyond Numeracy*

Pullman, B. 1998, *The Atom in the History of Human Thought*

Chapter 5. Believe or Know

In "The Search for Truth," we considered the difficulties inherent in the establishment of human truth. We asked the questions: what is truth; how can we know the truth; what are the steps to learn the truth; and finally, what really do we know? The present chapter examines further the idea of human truth and, in particular, the distinguishing characteristics between the processes of believing and knowing — subjective belief and objective knowledge.

To believe or to know, that is the question. First we must recognize that the truths we know or believe are unique; they are human truths about our sense and feeling of reality, not some kind of edict about objective, eternal verities. Human truth is always incomplete because of sensory and reasoning limitations. We often use the words believing and knowing as identical in meaning. That is not the case. From *Webster's Unabridged Dictionary* (1997) by definition, to believe is the opinion or conviction that something is true but not necessarily because of objective proof, direct observation, or experience. To know is to have knowledge of, or a clear and certain perception of a fact, where a fact is understood to be that which can be observed. To believe is to have faith in the truth of a statement, written or oral; to know is, in addition, to have confirming empirical facts about a statement. Of course, there are nuances, usages, and ambiguities that tend to obscure the distinction; however, to know must be based on objective observation of fact, whereas believing can be based on subjective mental feeling alone.

On Uncovering Truth

We can believe an observed fact as well as know it. That is probably the source of our ambivalence and confusion about the two meanings. However, in conformity with our experience, objective facts must always be considered tentative and subject to modification or elimination as new facts are uncovered. Beliefs are often held more firmly because there are no confusing facts to consider.

There are many disagreements as to what is true and what is false, but there is also little agreement as to what truth necessarily is. Again, we must recognize that our mental models of reality are simply that — models. There is no certainty that tomorrow we will not discover a better model. To find answers to these problems we first traced the history of human efforts to discover and confirm truth. We then discussed some methods that could decide the truth of an assertion, including deductive and inductive logic, trial-and-error, scientific, axiomatic, observational, and absolute procedures. The common threads working through the various methods are the themes: formulate the question, search for facts, compare, select, test and accept if worthy. However, the absolute method has nothing to compare or test against and; therefore, can have no observable proof. It must simply be believed; it cannot be known.

To evaluate the various methods for finding truth, first formulate the question to be asked and answered; second, search for all possible answers; third, compare each explanation with all others; fourth, select the most logical and fact-based explanation; fifth, test the selection, and sixth, accept the best tested selection as the truest for that moment. In actual fact, all six steps are often going on at the same time. If no explanation has convincing explanatory power, we may go back to additional objective observation and trial and error, or simply wait for new evidence. Since we never know when a better explanation will come along, our superlatives "best and truest" above, must be scaled back to the comparative truer truth. That is what we humans know and can ever expect to know. We can know only the better truths we observe at a particular moment in time. Remember that since all things are changing, constant reconfirmation is required. The method of confirming truths is a recommended sequence of actions to take when trying to uncover facts or the truth, not recommendations as to what to believe or think. Any event can be explored by the process of confirming truth; but heed this warning, there can be no guarantee of a preplanned or expected conclusion, and the truth of the moment might not endure for long.

We should note there are different kinds of truth and we often do not differentiate between them. Our discussion mainly concerns objective truth, which relates to observation of the external world. Subjective truth has different characteristics; it relates to those feelings and thoughts that go on within our bodies and brains. When you have a headache, you are certain that is a fact; you can observe it; what more proof is needed? If you tell a friend you have a headache, is your friend as sure of this as you are? Perhaps yes — she has complete faith in you and knows of no reason for you to lie; perhaps no — she has reason to believe you have a motive to lie — you don't really want to go to the movies tonight. The point is that she has no observable facts to go on, only her subjective feeling. In any event, her dilemma shows again how elusive it can be to find the objective truth.

The problem with subjective belief is that there is often only one person who knows the truth or falsity of a statement by personal experience; all others can only believe it. With objective observation many people can directly observe the truth or falsity of a proposed statement and come to a conclusion as to its worth. We might call this the democratic approach to truth since everyone has the right and often the duty to make observations, confirming or not.

The Nature of Knowing

The major problems with knowing are the adversity it meets, the time it takes, and the number or weight of required observations. However, knowing often prospers under attack. There is no better way to establish or confirm knowledge than to defend it against the bitterest criticism. The most powerful criticism of all is one formulated in such a way that it cannot be tested or observed.

For example, recall the EPR (Einstein, Podolsky and Rosen) group's objection to a part of the Quantum Theory described in the chapter Scientific Reality, or Lord Kelvin's criticism that Darwin's evolutionary theory was false because it required too much time, discussed in *The Double Helix*. The EPR objection was raised in 1935 and finally disproved by A. Aspect and his colleagues in 1982, after improved experimental techniques had been developed. Kelvin's objections were raised in 1854 and not refuted decisively until 1903–1931 after radioactive materials had been discovered and confirmed to produce heat and have accurate time-measuring capabilities. During these long periods of 47 and 77 years between criticism and final rebuttal, scientists were uncertain about what they could know or only believe. Were Kelvin and the EPR group correct or the evolutionary and quantum mechanics theories? Time and effort were wasted trying to confirm or disprove the criticisms. However, in their work on such problems, scientists sometimes learned more about other aspects of nature.

The time needed to clearly define a truth by observing and testing experimental data is universally more a problem than adversity. Criticism has always been an inherent part of the scientific process whether it was only trial and error or the more formal method.

The systematic planning and gathering of experimental observations for the purpose of understanding nature had little momentum until the modern era. The early scientists were usually at universities or one of the imperial courts of Europe. Many of them were employed on the basis of supplying their own costly experimental apparatus. At first they were barely tolerated among the professors of classic subjects such as theology, rhetoric, philosophy, law, language, and medicine. Science professors were often alone and were expected to be expert in many subjects — geography, anthropology, geology, physics, chemistry, mathematics, astronomy, and biology. Nevertheless, the universities slowly expanded their science faculties, if only for the prestige (Jungnickel and McCormmach, 1986).

It was not until the discoveries of science bore fruit in the technological development of practical materials and devices — the barometer (1643), hydrochloric acid (1648), steam piston pumps (1690), quadrants (1731), shuttle looms (1733), etc., that the full potential of scientific research began to be understood.

As nation-states came to appreciate the advantages of technically developed military armaments, the scientific enterprise received additional support that has steadily increased into the present day. After the destruction of the American Civil war that signaled a revolution in the art of war, an arms race ensued leading directly to the World Wars and the destruction of countless cities and slaughter of tens of millions of people. Most of this was a result of basic research and technological development of new and more powerful weapons.

We are now at a point where state funded research is primarily oriented toward military purposes rather than for the benefit of all citizens. Semiconductors, computers, software, nuclear power, radar, avionics, intercontinental missiles, atom bombs, internets, global positioning, and satellite communication are all heavily indebted to military funded research and technological development. Fortunately, these often have general applications to ordinary human needs as well.

In addition to the time it takes to accumulate large amounts of experimental data, there is the time necessary for a young scientist to achieve a competent understanding of the latest scientific advances in his or her field. It normally takes four years to obtain a bachelor's degree and another five to seven years for a doctorate in a scientific discipline. After that, the young scientist often has to spend years teaching and preparing to work on original research. Most advanced research is carried out at universities where it is funded by governments.

There is also the amount of data that is required to confirm a theory. When von Laue sent an x-ray beam through a crystal of copper sulfate in 1912, he was opening a vast new era of research that could determine the position of atoms in crystals. Recall, there are 92 naturally occurring elements, many of which can react with each other to form crystals in their solid state. There are also seven basically different ways that atoms can be arranged within a crystal cell — cubic, tetragonal, orthorhombic, rhombohedral, hexagonal, monoclinic and triclinic. Combining these variations and their permutations, we end up with a vast number of different crystal structures each of which can be identified by a unique x-ray pattern.

Now that we know some facts, namely that x-rays passed through a crystal can produce a unique pattern on a photographic film, how many times must it be repeated before it is considered reliable and gives consistent results?

Will two do, ten, a thousand, or a million? When a scientist gets two — he or she is encouraged, at ten — ecstatic, with different investigators and equipment — positively certain. To date, hundreds of millions of x-ray diffraction measurements have been made and they all agree with Bragg's original equation of 1913 (n lambda = 2d sin theta). That is the perfect ending. Probably most scientists are content with ten or twenty confirmations and then depend on other laboratories and investigators to anoint their work with final approval. All of this takes time, often lots of it.

We can compare the certainty of the scientific method of determining truth with some other methods such as political orations or corporate board meetings. Consider our judicial system, in which twelve jurors, two lawyers and one judge determine the truth or falsity of a legal charge. Often such procedures are determined by a single witness or piece of information determined by inferences which cannot be tested adequately. That would appear to be a light-weight method of determining truth yet we send people to their deaths on such grounds. Finally, when we come to subjective explanations or beliefs, we find they are based on feelings alone.

However, some will point out that a scientist cannot possibly have direct observation over every procedure or piece of data involved in carrying out an experiment or forming a scientific conceptual framework. It is true that scientists accept many authorities to collect and confirm data for them. That is why they are always critiquing each other's results at scientific conferences. If you visit the science section of a university library, you will see endless rows of bound volumes filled with reports on decades and centuries of scientific effort. Scientists who have an idea that elaborates on a scientific paper in an earlier volume will refute the prior data if it does not agree with their own. This is the cleansing action that constantly confirms or denies new data.

Nowadays, professors often use textbooks of a thousand pages or more to introduce students to a first year of sciences such as chemistry, biology or physics. How can a professor convince students that all this information is true? Several afternoons in the laboratory each week testing the factual content of the textbook and the professor's lectures, often convince the student that these are observable facts that form a harmonious whole. Nevertheless, a person must take the most cynical stance possible to guarantee that carefully selected authorities are accurate, unambiguous, and thorough.

As the student works his or her way through the concepts and observations of a science, a grand vision of nature in one of its many manifestations begins to emerge. If one studies several different sciences, the more encompassing, awesome, and overwhelming the vision becomes. However, for the general reader, the most convincing daily proof of the truth of scientific observations comes from the technical developments of science to produce new and useful products ranging from steam engines and machine guns to cell phones, autos, computer games, and medical drugs. The overriding motivation of technology and development engineers is to use the basic findings of science to achieve useful results. In this, they have been remarkably successful.

There is another worthy motive for seeking knowledge — curiosity about the universe we live in. Our curiosity leads us to puzzle over and explore the unknown. All creatures, even birds do it; after all, there might be something good to eat out there somewhere. By gaining knowledge, we might increase understanding and control over a desirable process. Also, one has the fear that a lack of knowledge might endanger self or kin.

For the individual citizen the message is clear; we must select with the greatest care the authorities we believe and constantly reevaluate them. We must continually update our general knowledge in order to understand more about the universe and who and what we are (Graziano, 1989).

The Scientific Method

We have referred to the scientific method often but not always to a satisfactory conclusion. The method does not end with the originating laboratory and its data. It ends with the publication of the detailed results of the investigation and the duplication of the results by other laboratories with different but competent groups of investigators. If other groups of scientists cannot reproduce the data, the original data cannot be accepted into confirmed main stream science. For clarity, listed below are nine major daily activities that investigators in chemistry, physics, and biology, etc., often follow:

1. The scientist(s) should become competent in the subject matter of the scientific field investigated. He or she must begin a continuing literature study of all related scientific papers.

2. Related papers should be collected and inductively compared for inferences or implications that might yield insight into new areas of knowledge.

3. Potential data from the new knowledge should be rearranged and/or modified to create a new hypothesis that could be experimentally confirmed or disproved.

4. Assemble the appropriate experimental equipment.

5. Experimentally test the implied and inferred facts of the hypothesis. Reproduce and retest the facts until there is no doubt that the experiments can be confirmed repeatedly. Keep detailed bound notebooks on all data and modifications of apparatus and procedure.

6. Deduce from the confirming data possible new insights and hypotheses that might also be tested and confirmed.

7. Write and send a detailed report of the data, equipment, methods, and results to a scientific journal for peer review and publication.

8. Give papers at scientific conferences that specialize in allied specialties. Encourage other laboratories to reproduce or extend the findings.

9. When independent researchers at other laboratories reproduce and confirm the data it can be accepted as part of the main body of science as long as no falsifying data appear.

The first three activities mainly involve study, searching, evaluation, and analysis to create a possible new hypothesis.

They determine the initial direction of the investigation. Aspects of the first six activities are normally being reevaluated and reworked throughout the investigation. The forth activity is a simple declarative sentence that describes what can become years of work among innumerable changes in direction. The fifth and sixth are the wearying heart of the matter: experiment, try, try again, repeat, tweak the equipment, try, and try again, and again. The seventh, writing a scientific paper, will send the researcher back to his equipment to confirm the confirming data. The eighth is sobering because the author, under questioning, often has to disclose his dumb blunders and mistakes. And ninth, after all that, he has to wait for some other group to reproduce his data and prove that he was honest!

There are sciences such as astronomy that are more observational than experimental. These manipulate the results of observations to infer the genesis and evolution of cosmic bodies and their movements. They apply the known principles of earth-bound chemistry and physics to the light coming from distant stars and galaxies. The light spectra from atoms on Earth, for instance, are identical to those coming from the stars; thereby inferring that the atoms in stars are the same as those on Earth.

We should note that the scientific method includes a temptation toward bias. A researcher spends every waking hour trying to prove the truth of his or her hypothesis; he is not trying to prove the falsity of it. He is always tweaking his experiments to prove he is right; not that he is wrong. Consequently, in experiments where there are only slight differences between confirming and falsifying data, there is a danger of treating falsifying data as due to random errors or mistakes in procedure, not as negative results. This is the age old conflict-of-interest problem that plagues most human activities.

The Dangers of Scientific Fraud

The slanted data described above are errors in procedural methods not necessarily fraud and are usually caught and corrected by the researcher or his colleagues. However, in recent years there has been increasing publicity about scientific fraud. The most recent example (2005) is the questionable results on stem cell cloning research performed by a South Korean researcher, Dr. Hwang Woo Suk and his associates. In the midst of the scandal, the charge was made by some that this proved the scientific method was broken and could not be trusted. In fact, the exact opposite was proved; the questionable results of the research were disclosed by some of the researchers, the publishing journal Science, and South Korean government investigations. The major problem was the overwhelming publicity given the research before an independent research group in an independent laboratory had confirmed Dr. Hwang Suk's results. Recall the ninth activity in our list above of requirements for the complete scientific method.

The history of science is strewn with scientific fraud, incompetence, mistakes, pranks, inflexibility, and ignorance.

The classic example of the above is the case of the Piltdown Man. Near the town of Lewes, England in 1912, Charles Dawson, a lawyer and amateur supplier of fossils to the British Museum found a skull and mandible that displayed primitive non-human characteristics. The fossils were found in an early Pleistocene deposit that was known as the Piltdown site. But the public was ripe for startling news since Neanderthal fossils had been discovered in Europe in 1856 and the fossil hunting world in England was in a fever for more fossils that could be missing links related to humans.

There were many skeptics of the Piltdown story but in 1917 a second skull was found that convinced most of them. Over the next few decades discoveries of fossils such as Australopithecus began to erode the importance of Piltdown man, which did not fit well into the line of new discoveries. In 1949, fluoride dating showed the fossils were not old and a more careful study of the Piltdown fossils proved them to be forgeries in 1953. The skull was from a modern man and the lower mandible from a modern orangutan with teeth carefully filed down to resemble those of a modern human.

The Piltdown fossils were undoubtedly a fraud or perhaps a prank but the perpetrator(s) were never identified. Of the five or six prime suspects, Sir Arthur Conan Coyle, the famous creator of detective Sherlock Holmes and Dr. Watson, would be a choice candidate. Coyle lived only seven miles from the Piltdown site; he was trained as a medical doctor and thus knew human anatomy; he had published a detective story, "The Lost World," in 1912, which had similarities to the Piltdown story; and was well acquainted with archeological professionals who could have supplied him with the human skull and orangutan jaw bone.

Regardless of the details of the fraud, the scientific world was burdened by it from 1912 to 1953 before the scientific method for determining objective truth was able to refute the Piltdown claims conclusively.

A number of reasons have been proposed for the recent surge in scientific fraud (New York Times, December 20, 2005). First, there is a global expansion of scientific research due to the diffusion of Western knowledge of science and technology into the so-called Third World. Japan was the first and most spectacular convert to Western science during the Meiji Restoration of 1868. Today, the giants India and China are following Japan's example, with populations that are much larger, and the rest of the world is close behind.

Here we must pause for a moment to distinguish between scientific research and technology, which use essentially the same methods. Scientific research, sometimes called basic research, investigates the basic characteristics of the universe to increase our knowledge and understanding; its questions are; what is it, what does it do, where does it come from, what does it weigh, and on and on. Technology tries to make something useful; its questions are: how can I make a profitable new drug, new hat, new motor; how can I make a paint that hides and protects a surface with a cheaper thinner coating; how can I implant a pacemaker that will make a heart beat properly? Basic research creates the core of scientific understanding; technology and engineering uses science to create goods and services for humans.

The profit motive of technology is often behind scientific fraud. The public assertions by tobacco company executives in the late 1950s that smoking was not the cause of lung cancer are a typical example of the collusion between technology and corporate policy to befuddle customers. The recalls of drugs, artificial heart valves, pacemakers, defibrillators etc. are more recent examples.

Another factor that fuels the upsurge in scientific fraud is the vast increase in scientific journals to 54,000 worldwide. Foreign journals in particular have increased from 15,300 in 1980 to 29,098 in 2005. The inherent problem is whether the increase in scientific articles can be given adequate peer review.

Although the profit possibilities of technology are a permanent invitation to fraud, personal acclaim also seems to be a motive. In 2002 a researcher at Bell Laboratories was fired for publishing seven papers in the journal *Science* and five in *Nature* using fabricated data on spectacular new electronic devices. In 1999, a group at the Lawrence Berkeley laboratories in California announced that they had produced in the laboratory new extremely heavy atomic elements with the atomic numbers 116 and 118. After much public acclaim, the results were withdrawn and a physicist was fired for falsifying the data in 2002. Even basic research appears to be succumbing to the allures of money and fame.

These perversions of scientific culture, faith, and method cannot be excused or justified for any reason. The heart of science is an unswerving allegiance to the search for observable objective truth regardless of the consequences. Anything less is a corruption that would leave humanity without a sustainable direction or future.

The Nature of Believing

The true believer has an entirely different set of problems and goals. To believe without confirming facts is limiting because alternate possibilities are not thoroughly considered or tested. The believer will contest the above statement and say: "I compare a new belief with my overall belief systems; if it agrees I accept it; if not I reject it." In other words, I accept only what fits into my earlier beliefs. This is the only test the new belief must pass, which is clearly no test at all. The scientists, in contrast, must constantly test and confirm their data, in detail, personally.

There are other motives that make a belief system founded on faith alone acceptable so that testing is avoided. It is simple and relatively easy to understand; it avoids doubt and conflict; ignorance; fear of learning something different; and is comforting. We can believe anything we want to. Belief often creates absolutes that can never change and are never tested; belief is the end of searching, curiosity, reasoning, doubt, questioning, and growth. There is a basic conflict of interest in the process. We often believe because it makes us feel good, not because we have honestly searched for or been willing to recognize an alternate objective truth (Shermer, M., 2000).

If you are free of beliefs, you can more easily accept new beliefs or knowledge. Beliefs can be addictive, inhibit action, and chain you to the past. Dominated by unchangeable belief systems, we live in illusions that endanger self, family, and humanity. If you believe what you are not, you will never know who or what you really are. The freedom to know means to let belief systems waste away as truer facts become available.

However, there is cause for optimism, living creatures have always found ways to justify, protect, and extend their existence. Observable facts are simply there to be accepted or not; they cannot be destroyed or denied by either belief or disbelief, but can be confirmed or denied by repeated testing. There lies our safety and our direction.

Motives for Belief

Since beliefs often cannot be tested, they need little effort. They are often only what we want to be true. The belief of others does not make them more or less true. We often hear what we want to hear, not what there is to hear. Belief can comfort us, but it cannot justify our actions. Beliefs are often a shield against failure, harm, shame, etc. Beliefs should at least be reexamined when they lack confirmation by objective observation.

Separating Belief and Knowledge

Our mental processes and language do not differentiate automatically between subjective beliefs and beliefs with confirming observations and facts. How can we be certain when we craft a simple declarative sentence that it will contain truth? The straightforward answer is that we cannot without effort.

How can we recognize and separate subjective beliefs from objective beliefs of knowing? The task is not easy because of the many memories beginning in childhood and continuing through adulthood that are not identified as to truth content or category. Begin by writing down a list of major beliefs that can be identified. This list can then be separated into two: subjective beliefs and beliefs with objective confirmation. Of course, this leads immediately to the question: what is, and how much confirmation is needed.

Considering the quantity of confirmation needed, we are led directly back to the dilemma raised in discussing the scientific method. Since there is no number or weight we can specify with certainty, we must depend on personal judgment. This can be dangerous ground. However, if we earnestly and honestly try to consider objectively the distinctions, we should be able to compile the two lists or at least learn something about our commitment to subjective beliefs or objective knowledge.

Bibliography

Foglin, R. J. 1987, *Understanding Arguments* 3rd Ed.

Graziano, A. and Raulin, M. 1989, *Research Methods*

Jungnickel & McCormmach 1986, *The Now Mighty Theoretical Physics*

Leplin, J. Ed 1984, *Scientific Realism*

Shermer, M. 2000, *How We Believe*

Webster's Unabridged Dictionary of the English Language, 1997

Chapter 6. Time and Events

Whenever we think of time we also think of changing matter, events and things. Whether it is the fall of sand through an hour glass, the passage of the sun across the sky, the swing of a pendulum, the tick of a clock, or the flight of an electron in an atom, we are always thinking of matter in some process of change. We should also consider the direction of time. Why does time go like an arrow in one direction only, from past to present to future? What is its source and where does it go? The simplest answer is that time goes nowhere; it is not like an arrow; it is a concept we construct from changing material events. Matter, which is mass and energy, exists. The interaction of mass and energy creates events and change. Time is a derivative of matter as it changes. Any other interpretation leads us to a universe with two major concurrent motions; matter in the process of change and time flowing in another channel. Change is a presentation of differing events; the sequence of these differences is time. Changing events create time. Matter is what changes; it is unstable as it flows from states of high energy to those of lower energy in accord with the observations and laws of thermodynamics.

Some Views of Time

The past is contained in each present moment. Over the centuries, the idea of time found many forms. Some think time flows in cyclic swings, others in a linear fashion. As we shall see, Newton's time is absolute, Leibniz' sequential, and Einstein's relative. However, Bishop Berkeley and J.M.E. McTaggart, together with the

Hindu prophets of Maya, claim time is an illusion; there are no things that change; therefore, there is no time.

Time measured by nearby clocks is called local time. This is the time we live in. Quantum mechanics, the science of the extremely small, describes the quantum nature of light as dual; it can act as a particle or wave depending on how it is measured. These wave/particles are quanta, the smallest possible energy changes; they are entangled in a non-local arrangement. The Heisenberg Uncertainty Principle indicates that certain paired properties of very small particles such as position and motion or energy and time, cannot both be known accurately at the same time. Since time is a part of any measurement of motion and change, the Uncertainty Principle brings a kind of fuzziness to our knowledge of time.

We live neither in a micro world nor far cosmic places. We are from a middle way, the macro way, somewhere between the atom and the earth. Life evolved here on Earth; we evolved here. Quantum mechanical or relativity phenomena are too small to distort the evolutionary levels of our experience and existence. Our atoms do conform to quantum principles since they are involved in the structure of atoms and the chemical bonding of molecules but above these levels, we are not directly affected. We are creatures of a benign earth that permits the innumerable reactions of carbon to form more and higher organic structures of increasing complexity. These processes only function at energy levels within a few hundred degrees of 20 degrees centigrade. If temperatures had varied much more than that over the last 3.8 billion years, we would not be here.

Of the writers who discuss time, St. Augustine was fascinated and wrote extensively about it. In *The Confessions*, written AD 397–401, he pleads for an explanation of time and describes the mystery of time as follows (book XI, 17, book XII):

> What then is time? If no one asks me, I know; if I wish to explain it to one that asketh, I know not; yet I say boldly that I know that, if nothing passed away, time past were not; and if nothing were coming, a time to come were not; and if nothing were, time present were not. Those two times then, past and to come, how are they, seeing the past now is not, and that to come is not yet? But the present, should it always be present and never pass into time past, verily it should not be time, but eternity.

More recently, Newton had different ideas. There is a famous passage from his definitions in the beginning of *The Principles* (Newton, 1687) that reads:

> Absolute, true and mathematical time, of itself, and from its own nature, flows equably without relation to anything external, and by another name is called duration: relative, apparent, and common time, is some sensible and external (whether accurate or unequable) measure of duration by the

> means of motion, which is commonly used instead of true time; such as an hour, a day, a month, a year.

Leibniz' third letter to Clarke in 1716, really meant for Newton (Alexander, 1984), explained Leibniz' thoughts on time as follows:

> But this itself proves that instants apart from things are nothing, and that they only consist in the successive order of things; and if this remains the same, the one of the two states (for instance that in which the Creation was imagined to have occurred a year earlier) would be nowise different and could not be distinguished from the other which now exists.

Einstein decided time is what clocks measure (Barnett, 1948) such that:

> The experiences of an individual appear to us arranged in a series of events... I can, indeed, associate numbers with the events, in such a way that a greater number is associated with the later event than with an earlier one. This association I can define by means of a clock by comparing the order of events furnished by the clock with the order of the given series of events. We understand by a clock something that provides a series of equal events that can be counted.

Bertrand Russell (1929) described a world in which only events exist and time as a human relational creation.

> We cannot point to a time itself, but only to some event occurring at that time. There is therefore no reason in experience to suppose that there are times as opposed to events: the events, ordered by the relations of simultaneity and succession, are all that experience provides.

St. Augustine eloquently defines the problem of time and things in the throes of change. Newton at first defines true, "absolute" time that flows and has duration. He compares that in the second part of his sentence to relative, common time; a motion that can be sliced into hours, days, and years. Einstein observes that a device that creates uniform, repetitive events (a clock) can be used to count out a measure of time and the relative sequence of other events. Russell dismisses time as derivative of events and Leibniz does much the same. All five descriptions have some parts in common, although Newton's absolute time has fallen by the wayside. When considered, they describe a world in which things (or events) exist, endure, and change, and that a sequence of equal events in the process of change (a clock) can be used to yield a number sequence called time.

But we should not forget the Hindu prophets of Maya and the Berkley, and McTaggart claims described in the second paragraph. They believe neither time nor change exists. And even Einstein, in a consoling letter to the grieving widow of his friend, M. Besso, made the strange comment: "For we convinced physicists, the distinction between past, present, and future is only an illusion, however persistent..."

(Conveney and Highfield, 1992). Whether Einstein was referring to the timelessness of the laws of physics or to a religious conviction is not clear.

Einstein's remark about clocks also requires amplification. Deductions from his relativity equations indicate that clocks of any kind slow down when they are moving or in a different gravitational field relative to the observer. These predictions have been confirmed repeatedly by experiments on mu mesons, gravitational field gradients, and clocks that are moving (in an airplane for instance) relative to a clock on earth. What Einstein was saying in effect, was that time undergoes changes that clock structures undergo.

Our ideas conform most closely with the views of Leibniz, Einstein, and Russell that local time is one of our ways of characterizing the general properties of events — change and duration. In this view, time has no existence of its own; it is not a separate phenomena; it is a derivative of events, happenings, processes, and changes. It is not a thing that can flow or carry matter anywhere. Time is a deduction from events as they relate to all other events. Finally, time is past, present, or future only in its sequential relationship to events.

Do past events have real existence? By real existence we mean, material existence, something we can sense or know now. To avoid confusion, we can distinguish between different classes of events using the following rough scheme:

1. Event — a present event, also the general term for all events.
2. Pevent — a past event — history.
3. Revent — a remembered event — an active class of pevents, DNA.
4. Fevent — a future event.

Events and Existence

All change occurs in the present, that is, in events. Events are matter in the process of change. These are the domains of action, instability, and change. The past, the region of pevents, stands complete and tranquil like a shadow of earlier events. The past is motionless, but seemingly receding from us. Pevents are the final, unchanging record of past happenings; they are the stuff of history.

The interaction between a photon and an electron can occur as a single event. But most events are processes with many discrete parts each of which can also be an event. There are limits, however, to our ability to resolve events into parts. We also have a tendency to lump simple events together in order to study their function as larger entities. Even an event such as a birthday party, is a very involved affair with many subordinate events going on at the same time. When an event is complete, it has transferred its residual characteristics to a new entity a pevent, and a fevent.

With a new present we come to another existence. All things are at a full level of being only at the present moment. Underlying past events are basic to present being and change. Our bodies, that remarkable complex of skin, bones, and organs, are maintained by innumerable chemical changes and at lower levels by protons, electrons, neutrons, quarks, etc. How much further down we can go with our observations is problematic. But that is of little evolutionary importance since at our ordinary level of life the quantum activity of subatomic particles is hardly important. No species became extinct due to the destructive interference of two photons or our uncertainty about the position of an electron in a cell. At each level of existence lower levels are less important to the immediate potential of the present.

Present Events — Action and Change

Time measurement or at least the awareness of passing time is, for us, its basic nature. Our biological senses give us an internal feeling of time's flow; our mood changes; we get hungry or grow weary. We remember: we did that yesterday and this today. How do we measure the apparent passage of time? Our prehistoric ancestors had only large processes like the passage of the sun or phases of the moon to guide them. In the past two millennia some humans have used clocks. A clock provides a continuous series of standard, uniform events. Consider some of the clocks we have used over the centuries (Whitrow, 1972): beside the sun and moon we have the tides, sundials, heartbeats, hour glasses, mechanical clocks (about 1280–1300), clocks driven by pendulums (1657), tree rings, radioactive decay, red-shifted photons, vibrating crystals, and atoms. These all produce events that we can observe directly or with gauges or counters. The word event in this context means a happening, process or occurrence. Photons, thoughts, reacting atoms, a collision of galaxies, are all events. The tick of a pendulum clock is an event produced by a pallet releasing an escape wheel one notch. We count a series of these ticks to measure time between changes or events such as hunger pangs and our last meal. An event is matter in the process of change; it is changing existence.

Although most clocks are based on cyclic events such as the swing of a pendulum, radioactive decay entails a nonreversing linear breakup of atoms into other atoms, particles or photons. The rate of radioactive decay is extremely slow in some elements. So slow, in fact, that only one half the uranium atoms crystallized in mineral specimens formed 4.5 billion years ago have decayed into lead and helium by today.

Since the decay rate for each kind of elemental atom is fixed, we can use radioactive decay to make measurements of time like a clock.

Any observable, regular, cyclic or linearly changing event can be used to measure time. Can we point to anything and say, there is time? No, there is not; there are only events and sequences of events. Time is a relational concept we use to help us understand the basic nature of our lives and the universe — change. Time is incorporated in change. Before the First Event, the First Change or the Big Bang, there were no events and no time (Hawking, 1988). Time is measured by events; events are measured by differences in matter before and after a change.

We define events. To a child, nature appears first as a continuum, a changing display of color and form that is incomprehensible. Slowly we can discern variations that can be isolated, remembered, and defined. The combination of our observation and the thing observed is itself an event. According to one atomic model, an atom is a ball of electrons distributed in shells around a tiny nucleus of protons and neutrons. All these particles are in ceaseless motion. Even though atoms are theoretical constructs, we can consider them events. An atom is an event because we can observe it changing, describe it, visualize it, and measure its characteristics. By focusing our attention on differences, we break nature into an endless variety and array of events. We decide when and where an event begins, ends, and what is in between, according to our reason and to suit our purposes.

Events come in every size, shape and arrangement. Although we see them as separate, they are often interrelated by cause and effect and cannot always be separated. Among the simplest events are interactions between particles and photons such as, electrons and light rays. Photons are in unending motion, moving at the speed of light in free space (186,000 miles per second). They have been moving since the Big Bang, or First Event, unless intercepted and absorbed by other matter or radiated from more recent atomic events. Particles such as electrons and neutrons move at slower speeds. With the atom we come to a still higher level of complexity. Each atom has many events occurring within its orbits and nucleus, as well as new kinds of events that involve it as a unit. For example, atoms can unite with other atoms to form molecules, clouds or stars. With life the process continues: molecules combine to form cells; cells assemble to form organs; organs combine to form living creatures; creatures form groups or organizations. Each of these levels of complexity produces new kinds of events that contain numberless events at lower levels. All are intertwined, entangled, and interdependent at both higher and lower levels. When we look at a chair, we see no atoms or electrons although we know they are there. We see a shape, color, position, mahogany, a place to sit, the décor of a room, but no atoms. Using a different set of observational equipment (x-ray tubes or balances, for instance) we would observe other kinds of events.

The store of scientific information has been growing at an accelerating pace over the last three centuries, but only recently has it been possible to place much of it in coherent time and place sequences. The major events leading to these advances were Hubble's proof of an expanding Universe in 1929 and Becquerel's discovery of radioactivity in 1896 with final confirmation of radiometric dating techniques by 1931. The experimental tools developed during this period gave scientists the means to measure remote events in the history of the earth and the universe. With these, the distance to far galaxies and the time to past eras became measurable.

Not only were the radioactive elements considered a source of heat, but evidence was also accumulating that the radioactive decay process could be used directly to determine the age of the earth. Over the next few decades many of the details of radiometric dating were established. In 1921 and again in 1922 international symposia on the subject determined 1.3 billion years to be the age of the earth. In 1926 the US National Research Council established a committee to prepare a comprehensive survey of the field. When the report was published in 1931 the radioactive decay of uranium to lead was cited as the most accurate dating method and gave 1.5 to 3 billion years for the age of the earth. The age of the earth has steadily increased as older rock strata were discovered and dated. The present age of the Earth is estimated to be 4.6 billion years from recent data.

Of the many radiometric dating systems now known, the uranium/lead system has been most widely studied and used as an earth dating method. Uranium 238 radioactively decays through sixteen intermediate steps ending in lead 206. The numbers 238 and 206 are mass numbers and refer to the number of neutrons plus protons, each equal to one mass unit, in the nuclei of the respective atoms. After eight alpha particles (helium nuclei) of mass 4 are emitted from a uranium atom of mass 238, the remaining mass drops to 206 and the atom is transformed into lead 206. It takes about 4.51 billion years for one half of a sample of uranium to decay into lead. This is naturally called the "half-life," a term used to characterize the time span of all radioactive atoms. Half-lives vary from fractions of a second to many years depending on the atomic species involved. Once the half-life has been determined by experimentation, the age of a rock can be calculated by measuring the amount of lead 206 and uranium 238 it contains.

The following tables are drawn from Margulis and Sagan, 1995, and Strickberger, 2000. They contain some important dates as determined by red shift and radiometric dating methods.

Universal Physical Events	
The universe began in the BIG BANG expansion	13.7 BYA*
Heavy atoms (C Si Pb etc.) created in supernova	11 BYA
The solar system evolved	4.6 BYA
The end of meteor showers on earth	3.8 BYA
Universal Biological Events	
Age of the earth	4.6 BYA
Oldest meteoric rocks	4.5 BYA
Earliest moon rocks	4.2 BYA
Origin of life, anaerobic prokaryotes (bacteria)	3.7–3.9 BYA
Earliest fossils — the Apex cherts	3.5 BYA
Oxygen abundant — aerobic bacteria	2.0 BYA
Bisexual reproduction, the eukaryotes	1.5 BYA
Cambrian era — animals, plants, fungi	500–570 MYA
Mammal-like reptiles	290 MYA
Largest extinction — Triassic	245 MYA
Cretaceous-Tertiary, dinosaurs extinct	65 MYA
Modern humans — *Homo sapiens*	0.2–1.8 MYA
Human cave painters	50 TYA
Last ice age	10 TYA
* BYA = billion years ago, M = million, T = thousand (estimated)	

The Ascent of Man

A scientific theory, philosophic analysis, or understanding must be compatible with these short lists of empirical facts if they are to achieve a significant level of confirmation. If the longer time estimates are in error by even a hundred million years, the event sequence would still hold and the error would be insignificant at a few percent. These observed events will undoubtedly be altered, expanded, refined and replaced as observational techniques continue to improve but their conceptual cores will remain for the foreseeable future. A total reversal of any one of them would be surprising but not fatal. A follow-on theory would simply explain the data more accurately as Einstein's relativity theory (1905) explained Mercury's orbital precession better than Newton's theory (1687). Science is true because it can change with truer data; it can also be treacherous because one never knows when a seemingly basic concept will be altered to accommodate new data. Since observations are growing and changing, the scientific literature is full of such tentative words as: perhaps, it seems, it may be, approximately, generally, usually, etc. etc. Thus it is that the anti-scientists are often faced with the appearance of new scientific data or theories at the very moment when they are celebrating for having slain an old one. We should always remember that theories are wonderful, fruitful and satisfying, but observed data are forever. Much of the data once used as proof of Newton's universe now support Einstein's universe.

As humans we are naturally interested in more information about the evolution of modern humans:

The Ascent of *Homo sapiens Homo sapiens*

Homo sapiens Homo sapiens	‹35 KYA science, moon shot etc
Homo sapiens neanderthalensis	150–35 KYA complex tools
Homo sapiens "archaic"	500–150 KYA complex tools
Homo erectus	1.5–.5 MYA fire, tools
Homo habilis	2–1.5 MYA early stone tools
Australopithecines	4–2 MYA bipedal
Hominid ancestors	8–5 MYA few fossils

Matter, Mass and Energy

The universe is made of matter and as far as we can observe nothing else. Matter has two quantifiable aspects, mass (M) and energy (E). These can be related through the equation $E=MC^2$, where C is the constant speed of light in a vacuum (3×10^{10} centimeters per second). The equation states in essence that a certain quantity of mass is proportional to a definite quantity of energy and vice versa. Either can be converted into a specific quantity of the other.

A quantity of mass is defined by Newton's First Law of Motion as the resistance an object has to a change of motion in a straight line or state of rest (this resistance is called the inertia of an object). It is the extent to which a mass cannot be deflected from its flight path or state of rest. Mass corresponds somewhat to our ideas of weight since mass and weight are proportional. The two can be related through the notion of mutual gravitational forces that act between all mass containing objects. However, weight depends on the relative mass of gravitationally interacting bodies, not inertia; we weigh more on earth then we do on the moon, for instance.

Through our familiarity with weight we get a practical feeling for mass (M) but the physical meaning of energy (E) is abstract and illusive. In the 1840s and 50s a consensus was reached among scientists that all the changing processes of nature conform to a basic law, the Law of Conservation of Energy or First Law of Thermodynamics. Since mass and energy are proportional and transformable, we can equate the two and include mass in the term energy. The law states simply that within a changing group of material objects (or system, in scientific parlance) the total amount of energy before a change is the same as after the change. Or equivalently, mass and energy can be neither created nor destroyed. The system must be isolated so that nothing can be added to or subtracted from it during the change. A system is any isolated mixture of mass and energy we can observe. In the universe there are many forms of energy including: mass, kinetic, mechanical, gravitational, radiant, heat, nuclear, potential, magnetic, electrical, and chemical energies. Under the condition of isolation, any material mixture will have the same amount of total energy after a change as before. We have never observed a deviation from the law of conservation of mass and energy.

In nature there are particles that we call antiparticles, which have characteristics opposite to each other. For instance, the electron has an antiparticle called a positron, which has a positive charge rather than a negative charge. If these two particles meet, they annihilate each other and change into a quantity of energy equivalent to their combined mass as predicted by the equation $E=MC^2$. These particles and their interactions have been observed in naturally occurring cosmic rays as well as in laboratories with particle accelerators. Before and after such reactions, the total energy of the system has remained constant.

After all these considerations, however, we still do not have a clear concept or feeling of what energy really is. It does not lend itself to a visual model like an atomic nucleus with electrons swirling around it. It tells us nothing about the mechanism or processes of change but total energy is always the same, before and after the change (Feynman 1989, E.B. 1989).

The Nature of the Past

As events unfold and fade, pevents are created. Pevents are devoid of the attributes of matter. They have a different character, which we will call pexistence. A pevent is never completely gone; it can still have influence here in the present; it is a memory of a once material presence; it is history. When we consider time and the sequence of events, we can say: present events are the most recent summary of past events; past events create the present; or now is the whole record of the past. Pevents are not things in the ordinary way; they are mass less, unchanging, eternal and usually beyond our abilities to detect.

As events change and pass, pevents remain. If there are events and change, there must be prior events. An event becomes a pevent, which no longer exists in terms of matter, but is somehow still part of the present. We have a very limited ability to detect or trace pevents in this state of previous existence or pexistence. That we know of innumerable pevents is beyond question. Events are factual; pevents are records of the fact and also facts — eternal facts. Pevents are without mass or energy and therefore stable; they cannot change. In the total universe of possible events, pevents are unique in that they once existed as matter. Present events are built on a foundation of pevents.

Buttressing the idea that pevents influence existence is our awareness of the power of information. Information has some of the characteristics of pevents; it is without mass or energy and is a result of events now past. Whether plans to build an atom bomb, DNA to create a worm, or scores to perform a Mozart opera, information is from the past. The information behind every scientific theory or philosophic principle comes from the accumulation and organization of past experience. Information is created in the present as an event. Once created, it becomes a pevent. Pevents are like a residue of events or the last scent of a cup of coffee.

We are certain the present exists; when it changes we are certain it has disappeared into what we call the past. The past is fixed; it cannot actively influence the present unless past traces exist in the present and can be detected. Information about the past is in the present, but our ability to trace it through the endless sequences of events is limited. We have only to remember the discussion on fossils. Fossilization is a very rare event, mainly restricted to the last 550 million years after the Cambrian explosion. Earlier life, representing about 85% of life's tenure on earth, has left few traces, no skeletons, bones, or shells. Does that mean there was no life prior to the Cambrian explosion? No, it means we are deficient in our present means of detecting prior events.

Pevents often have an informational character with a very human basis. For instance, there are events retold by Plutarch in the first century AD from histories available to him. His sources were pevents; they allowed Plutarch to tell us the history of an ancient Greek, Timoleon, an aristocrat of Corinth who saved his brother's life in battle against Argos. When his brother made himself tyrant of Corinth, however, Timoleon agreed to his assassination. From shame and remorse, Timoleon went into self-exile for the next twenty years. When Corinth's colony of Syracuse asked for help against their tyrant, Dionysius the Younger, the Corinthians appointed Timoleon to lead an expedition against the tyrant. After a series of battles, Timoleon defeated Dionysius and returned Syracuse to a republican form of government. He died about 338 BC as the most honored man in Syracuse and a man of historic integrity.

Plutarch's story, condensed from his *Parallel Lives*, is a pevent of the historic kind. Timoleon's life and acts still inform us, stir our admiration and interest; they are not completely gone. In this context history is used in a broad sense, including prehistoric as well as historic accounts. Archaeology, geology, and all other sciences that give us information about the past, fall into this category.

However, our sensory organs are so attuned to present events there is little time to cope with pevents. Our attention is toward the most recent sensory events; our survival depends on them. When an auto goes whizzing past, I no longer need concern myself; it is the one rushing directly at me that is dangerous. "What is past is past," as the saying goes.

There are a great variety of pevents that endure such as the expanding universe, the red shift of light, atoms, radioactive decay, spectra, geological strata, and so on.

The expansion of the universe has been going on for about 13.7 billion years. At each moment of that time, we could have said with truth, "The universe exists and was preceded by a large number of events." We can still say it but have no way of knowing, in the sense of detecting, each individual past event. As living creatures we evolved by protecting and maintaining our present lives not by knowing or understanding the past. The red shift of light we detect today might be due to the light emitted from a galaxy millions or billions of years ago. For instance, the red-shifted light from the giant elliptical galaxy in the Coma cluster of galaxies has been traveling at the speed of light for about 350 million years to reach us today. We know this because the wavelength of a spectral line of the element calcium from the galaxy is drastically increased. When it was emitted 350 million years ago, the light had a wavelength of 3933 angstrom and now on earth we measure it as 4018 angstrom. This difference in wavelength, factored into one of Hubble's equations, calculates as 350 million light years. Each moment of existence along that long path is now in the region of pevents. Our abstraction of information from such pevents demonstrates that past events can be inferred and known by humans. As we look at spectral data from the galaxy and from radiating calcium here on earth we participate in an event with a pevent. At every moment of existence we are enmeshed in past events.

But do past events have pexistence or past existence only if we can detect them? Every photon that leaves the elliptical galaxy in the Coma cluster radiates towards other worlds; we just happen to be in the path of a few of them. It is certain that photons from Coma were intercepted and seen by creatures on earth long before we arrived. Detected or undetected, the pevent photons from Coma are flooding the universe at this moment. They are pevents in the sense that they left Coma a long time ago and they have been cruising through space ever since. If we look at Coma today and then not again until tomorrow, do we have the right to deduce, that is, to know from observation that Coma photons were also arriving in the time between today and tomorrow? Human detection has little to do with pevents; we are simply fortunate that we can observe and know something of them. The past is largely unseen but we know very well that it is there in its entire splendor.

Revents, Remembered Events — DNA

With the creation and evolution of life, new and different kinds of pevents appear: fossils, DNA, memory, history, etc. Such pevents have special characteristics; through us, they can take an active role in the present. These are revents or remembered events, a subset of pevents.

Revents are carriers of information that exist and are observable in the present. As we move deeper into the earth and out into the cosmos, we find more fossils, more artifacts, more galaxies and more relationships. From these we gather new information. With new information we find that some pevents are really revents. If we were infinitely observant and wise, we might find all pevents are also revents; but we are not; we must make do with the little we observe and can reason about.

For life, the supreme and unique revent is DNA, deoxyribonucleic acid. DNA is the accumulated record of the outcome of natural selection over the past 3.8 billion years or so. If anyone asks your age again you can, in truth, respond: Oh, about 3.8 billion years. Not only is DNA a carrier of information about the additive results of variation and natural selection processes, it can also give instructions in the present to reproduce itself and take action in the present. It can assemble materials and call up chemical reactions that will reproduce another individual of its own kind or its host. DNA is really a higher subclass of revents because of its unique ability to act at the present time with information gathered from the past. DNA is a vast memory, a revent that can act independently now. It can act now because it is existing matter capable of self-directed informed change.

We might say atomic structure or radioactive decay is similar to DNA. An atom has an electronic structure that determines the reactions it can have with other atoms. The 92 naturally occurring elements also have accumulated histories; most of the heavier elements were created in the supernovae of about 11 BYA and more recently. Their structures determine the nature of their chemical bonds. But atoms do not seek out others nor can they reproduce themselves. Radioactive materials decay spontaneously into new kinds of atoms, particles and photons. But the decay is random; it is a probability; it also cannot reproduce itself. DNA is in a unique class of its own.

With the emergence of DNA, life in its revent form can create events and pevents at many levels. We ourselves are revents at a still more complex level than DNA. We eat, drink, survive, and reproduce at a higher level, but largely under the direction of DNA. Some argue that we are only the hosts, the containers, the survival machines, for DNA. We are designed by DNA to carry them along as protection from the hazards of the environment (Dawkins, 1976). We certainly do that; they make us and we make them. It is a symbiotic relationship; we both win. But we have the added characteristics of consciousness and reason. Although at the mercy of our DNA in many ways, we are at a more complex level of being and therefore can act in new ways.

Consciousness is like a sixth sense; it allows us to abstract information from our senses and reason about possible actions and formulate the consequences of each. In this way we seem to exist with many futures from which to choose. Each day we must choose one, consciously or unconsciously, before proceeding to the next.

Revents can project their influence or create events because they exist in material form now. Like the chromosome's DNA, they can be observed now; they can cause events to happen now. They are of two worlds; formed in the past they enter the present in material form and take action in it. Revents, in their operational form are events and also pevents.

What happens then when brain cells and DNA disintegrate and disappear, changing into other forms of matter and energy? The revents that can create events in the present are gone. A carbon atom, once part of a brain cell, diffuses out into the world perhaps as a molecule of carbon dioxide that is absorbed by a tree and becomes part of a branch. What remains? Pevents remain.

After losing material, molecular organization and information are also lost. The carbon atom that once participated in human information from the past is now part of a tree branch. As other changes of the carbon atom occur, the relation between pevents will become more tenuous and certainly far beyond our abilities to follow. These relationships must, nevertheless, still have some kind of remnant pexistence. There is a new entity: the fact that with a certain event, a carbon atom took part in bringing from the past the information in DNA from lives even further in the past. Pevents and revents are unique in that they were once associated with matter. Events are changing matter; fevents are without matter; pevents and revents were once embodied in matter.

The events of today create the universe of tomorrow. This new universe comes with the fourth kind of event — the fevent. The future is mysterious but it is not totally opaque; some scientific laws project far into the future as well as into the past. The ancients also knew a great deal about the future: the sun would appear again tomorrow; winter follows fall; death will come one day.

Earlier we considered Bragg's law, which explains that if a crystal surface is exposed to a beam of x-rays, the beam will be diffracted depending on the angle between the surface and the beam, the wavelength of the x-rays, and the spacing between atomic planes in the crystal or in scientific terms *n lamda = 2 d sin theta*. This is a prediction based on enormous amounts of experimental data. The relationship has never failed for crystals from the earth, the moon and even asteroids. The crystals can be from recent samples made in the laboratory or the most ancient rocks on earth dating from 4.5 billion years ago. There is no statement about time in Bragg's equation. But we know that crystals of the same chemical composition made yesterday, today, or tomorrow will show the same relationships between angles, wavelength, and interatomic distances. Since these comments hold in general for scientific laws, we have a great deal of predictive knowledge about future events as well as past events. We should note; however, that scientific laws are true only for specific conditions of composition, temperature, pressure, mechanical stress, gravity and on and on. These conditions are not explicitly stated in the equation itself. For instance, if crystals are heated, their x-ray patterns change because heat causes larger atomic vibrations; therefore, the interatomic distances between atoms change. Usually interatomic spacing increases with heating but there are rare cases where the spacing actually decreases.

Another example is the chemical reaction between oxygen and hydrogen gases to form water under the conditions present on earth. Water and some energy are always produced when these two gases are brought together in a specific ratio and environment. The result of this chemical reaction, under exactly the same conditions, will always be the same whether carried out here or in another galaxy.

If we lay out the main characteristics of events, pevents, revents and fevents, we see more clearly their differences. Events are changing matter; they are observable, unstable, incomplete, transient, and causative. They are in the process of becoming. Pevents are mass less, stable, unchanging and complete; they are passive through time. Revents are summaries of the past that can act in the present. They bring information from and about the past that can be used to create new events. In their creative form they have mass, structure, and potential. Fevents are without mass, potential, probable, virtual, and with no memory. But they are constrained by their prior event and its load of pevents.

Duration, Past and Future Events

Events in the present have precursors and results. All change occurs in the present. Once an event becomes a thing of the past it enters a mass less and changeless realm. If we try to generalize what we observe in nature, we see first, ceaseless change and second, continuation. These are apparent contradictions. However, change is not haphazard; an electron never changes into an egret, or a cup of tea into a sea. As an electron changes, for instance, it continues as a changed event, perhaps as part of an atom or current flow in a wire. These are only a short way along the sequence of relational change. Events are interlinked into chains or webs through the processes of cause and effect. If this were not so there would be no logical or scientific relationships; each change would reveal an unfamiliar new universe. At this moment there is an oak tree, the next it is a raindrop or a photon in the far reaches of space. Our scientific equations are basically statements about the changing quantitative relationships between contemporary events. Although each moment brings a new universe, it is only slightly new; it is still a universe with which we are generally familiar.

We think of ourselves as current, as beings of the present, as moving each moment from the fading past to the incipient future. By such incessant activity we remain fixed in a seemingly eternal present. In reality we are made of the past, of past events that have endured into the present moment. Endure, duration, they are the keys. Events do not occur instantaneously from nowhere; they endure for various periods relative to other events. In our substructure we change constantly (ninety eight percent of the atoms in our bodies are replaced every year), but above the atomic, at a higher level of being, we endure. For instance, our general body form endures for a lifetime even as it grows, matures, and ages. What we consider to be our central being, our self, personality, mind, soul, or whatever term we choose; is made of the past, DNA, past experience, education, physical growth, relationships with family, friends, community, and organizations. All these were formed in present moments now past; they are traces of the past brought into the present defining our beings. Combined with incoming sensory impressions and reason, the self becomes a new being but again only slightly new; the self at any moment is not readily distinguished from a previous moment. This is true except for those rare moments of illumination, mental collapse, or death.

In addition to past events, or pevents as we call them, and present events there is a subclass of pevents that can act or cause action in the present; examples are fossils, geological strata, information in general, memory, and DNA. These we call "remembered events," or revents. There are also future events (fevents) that we can know accurately. The mathematical equations of science tell us that the light from the sun will take about eight minutes to reach us now, tomorrow or a thousand years from now and that sodium and chlorine will react under conditions similar to Earth's to form sodium chloride until the end of time, and so on.

DNA is in a unique class by itself among remembered events or revents. It is the fossilized results of billions of individual lives over a span of almost four billion years. It contains information won from the clutches of natural selection by innumerable individual lives. This knowledge is exhibited in the ability of DNA to replicate itself and recreate its host — even us.

Dates in Time

Where does all this belaboring of time and events lead us? We know things happened in the past but until recently we knew little about when, where, or the sequence of dates for distant events. Within living memory, our abilities in the measurement of time have expanded immeasurably. The result has been a growing need for reinterpreting the meaning of past events and history by bringing them into conformity with the new dating techniques.

The first scientific time-measuring breakthroughs came with the discovery, confirmation, and utilization of radioactivity in the period 1896–1931 and the expanding universe concept of Hubble and his colleagues in 1929. Before that the ancients had little more than the passage of the seasons in a year and the sun in a day to guide them. After the discovery of writing there were also local histories available from about the second millennium BC.

The ancient prophets calculated their first specific dates from the Bible. They concluded that the universe was created about 4000 BC, or 6000 years ago. Bishop Ussher (1656), John Lightfoot (1644), and Martin Luther (1512) calculated about the same date using the same source. After Darwin proposed his theory of evolution in 1859, the need for definite dating of past events became crucial since the evolutionary process required long periods of time. Unfortunately, Lord Kelvin, a noted physicist of the time, "proved" in 1854 that the sun had heated the earth for no more than five hundred million years. In 1868, Kelvin delivered a polemic against long geologic time and by inference evolution. He calculated that the earth had been cool enough to support life for only the last twenty million years. Before his death in 1882, Darwin never had the satisfaction of knowing his theory's requirement for much longer time periods would be met. Radioactivity and the red shift of light would prove there was plenty of time for evolution to occur but that was not confirmed until 1931.

Consider now the most momentous event in the evolution of complex forms of life — the evolution of bisexuality in the time period around 1.5 billion years ago. Before that, living creatures consisted of a single cell that reproduced by binary division into two essentially identical individuals. With bisexuality, two specifically related individuals, male and female, had to cooperatively interact in order to reproduce. The result was the first organization, the family unit. Within the family, the young were created and sustained until capable of maintaining themselves. The loving care of the parents for their young was the first instance of ethical behavior among living creatures.

Bibliography

Alexander, H. ed. 1984, *The Leibniz–Clarke Correspondence*

Barnett, L. 1948, *The Universe and Dr. Einstein*

Conveney, P. and Highfield, R. 1992, *The Arrow of Time*

E. B., 1989, *Encyclopedia Britannica*

Hawking, S. W. 1988, *A Brief History of Time*

Margolis, L. and Sagan, D. 1975, *What Is Life?*

Newton, I. 1687, *Mathematical Principles of Natural Philosophy*

Plutarch c. AD 46–120, *Parallel Lives* (Dryden translation)

Russell, B. 1928, *Our Knowledge of the External World*

St. Augustine, AD 397–401, *The Confessions*

Whitrow, G. J. 1988, *Time in History*

Chapter 7. Things Change

The all inclusive, fundamental nature of the universe is change. Over the past few hundred years many theories defining physical and biological change have been advanced and confirmed. However, these have not greatly altered our historical view of the universe and ourselves. That things change is hardly a new idea but the pervasive, all inclusive changes in the universe are not generally recognized or accepted.

In our universe of particles and photons, all are moving in relation to all others. Nothing is at rest; nothing is changeless. Yet the implications of unending change have not been integrated into a coherent world outlook. In general, we can know and understand the better of two related concepts but we cannot know absolute truth because that is fixed, changeless, and cannot be compared. We know only by comparison. Present knowledge from comparable possible facts confirms the view that we evolved from simpler life forms, are made of atoms and are essentially alone in an expanding universe. Since there is little probability that we will find or communicate with other intelligent life, the fate of human life lies completely with us here on Earth. Long ago in about 500 BC, Buddha clearly told us in Dhammapada 160, "The self is master of the self: for who else could be the master?" Yet now, twenty-five hundred years later, have we accepted responsibility for our acts and our role as the arbiter of life on planet Earth?

If we can integrate these modern converging facts into a coherent whole, perhaps we can find a truer understanding of the way things are and how we might cope with problems such as pollution, extreme poverty, excessive population growth, war, unbridled wealth and power, inequalities, energy and material deficiencies, lack of freedom and so forth. As with most such efforts, here we can dwell on only a few problem areas. Hopefully, these few will be of value.

Humans have done many grand and wonderful things. We have built huge monuments to honor our dead, dug great ditches to connect the seas, and fashioned long walls for protection. But there are those who say the most enduring work of mankind has been the search for truth. Seldom found, often doubted or ignored, has truth still eluded us? Do we know absolute truth or relative truth or truer truth or any truth at all?

The basic characteristic of existence is change. But if everything is changing, how can we know what is true? How can we know that things exist and change? Over the centuries philosophers, priests, and prophets alike give differing views on such basic questions. Among the ancient Greeks, those who expressed themselves most beautifully and convincingly is Heraclites, who declared everything is always changing; you cannot step into the same stream twice. Zeno, in one of his paradoxes, replied that nothing changes; in a race, Achilles can never overtake a turtle that has been given a small head start. For each stride Achilles takes to overcome half the turtle's lead, the turtle has also gone forward a small distance; with Achilles' next stride, the turtle has gone forward again thus always retaining the lead. Evidently, the ancients of 445 BC never thought in terms of distance per time period or feet per minute for instance.

In the Far East we find Lao Tzu describing the Tao — the All Powerful; doing nothing it causes all things to be done; going nowhere, it goes everywhere. The Hindu prophets of Maya teach that all we see and know is illusion, there is nothing real out there; behind the illusion is the true reality, the Brahman/Atman, from which we come and soon return. And Buddha tells us that life is chiefly suffering caused by craving. Suppress craving and we can achieve Nirvana, an unchanging state of perfection, with peace, bliss, and eventless nothingness.

In more recent years Descartes proclaims thought as the basic key to existence (the famous *Cogito ergo sum* — I think, therefore I am). Can only thinking things exist? How about facts, bats, or worms? Bishop Berkeley is sure that all we really know are the images in our heads. They correspond to no external reality: when you look from a table to a tree, the table ceases to exist and a tree comes into existence. And J.M.E. McTaggart decided there is neither before nor after, no time or change. "Nothing really changes. Whenever we perceive anything in time... we perceive it more or less as it really is not."

Aristotle, in his *Physics*, discusses many of the ideas about change that agitated the ancient Greek intellectuals. He has these remarks in Bk. VIII, Chap.3, and Para. 2:

> To maintain that all things are at rest, and to disregard sense-perception in an attempt to show the theory to be reasonable, would be an instance of intellectual weakness: it would call in question a whole system, not a particular detail: moreover, it would be an attack not only on the physicist but on almost all sciences and all received opinions, since motion plays a part in all of them.

Then, in what we must regard as Aristotle's Great Straddle, he states in the next paragraph:

> The assertion that all things are in motion we may regard as equally false, though it is less subversive of physical science...

The universe is changing and changeless, moving and motionless! We should not be too critical of Aristotle, however, since he modified his position somewhat in later paragraphs but still without quite settling on a static or changing world. Evidently his thought is that some things are at rest and others in motion; it depends on where you sit or look.

In recent centuries science has uncovered a vast array of changing phenomena that cast doubt on previous views. Still, we have no absolute method to prove that things exist and change. Here "things" refers to events, happenings and occurrences that are made of or related to matter, i.e., mass and energy, time and space. To avoid this quagmire of existence and change from which philosophy has yet to emerge, we accept for the moment the common-sense view that there is no need of proof; existence and change are necessary and obvious to our senses, reason, and lives. Perhaps we should add that to a human, "proof of truth" is a hunch or feeling of comfort, certainty, consistency or testability; in human truth, there is nothing but feeling not yet chiseled in stone.

The view that observation, memory, and reason are all part of our ability to be aware of existence and change is the most reasonable. Without observation we have nothing to reason about or remember. Without memory we can recognize neither existence nor change. Without reason we can compare nothing; memory and senses alone would give only a chaotic sequence of transient sights and sounds. And if, following Descartes, the only thing we know is that we exist, we are like beings encapsulated in eternity, without sensory impressions, without facts, without ideas. We must walk on three legs to reach even the foothills of understanding; these are observation, memory, and reason.

Change in Motion

The ancient Greeks, as typified by Aristotle, distinguished four kinds of change: motion or movement, number, quality, and coming to be and passing away. Of these four, motion was considered most common and revealing. The passage of the sun across the sky reminded the Greeks every day of change. With only oil lamps or firelight, the moon, planets and stars at night were a much larger part of their experience than ours is today. Winds and ocean waves, the flow of the seasons, work in the ports and olive groves, the endless drill of the heavy armored hoplites, and the frequent walk to the communal well for water, all gave a common feeling of movement. But they never knew nor imagined the level of movement we experience today. Travel that requires thirty minutes might then need the whole day, or week. Messages that are delivered almost instantaneously today might require months by boat, foot, or horseback. Even these examples of accelerated motion pale in comparison with the ceaseless activity science now reveals as the pervasive characteristic of nature.

If we look down, into the particulate nature of things, into our hands for instance, we now know of cells and tissue with blood coursing through channels made of complex molecules moving in harmony with the prevailing temperature and pressure. Each molecule is made of chains of carbon and hydrogen interspersed with oxygen and nitrogen atoms. All are vibrating at their natural frequencies, all moving with respect to each other. Within the atoms are electrons layered in swirls of probability; protons and neutrons spin while a plethora of other particles go their appointed ways.

Solid matter on earth is usually in crystalline form. The core is partially molten, but the earth's crust on which we stand and live our lives is largely clay, sand, and rock. These are all crystalline. When atoms coalesce and chemically bond together, they usually arrange themselves in a repeating, three-dimensional array called crystalline. Molten materials that are rapidly cooled can sometimes form glasses, which have a less regular repeating pattern. But whether in crystal or glass the atoms are constantly vibrating around their equilibrium bond positions. Only at the zero of temperature (zero degrees Kelvin, or −273 degrees centigrade) will these motions stop.

The real world is made of matter in the form of mass and energy. Our bodies are made of matter; we are designed for it as houses are designed for wood or brick. Or more exactly our predecessors designed us for it. Matter is unstable; it is constantly changing. Wherever there is stability, we find an arrangement of matter that appears as an equilibrium condition. For instance, the solar system has had large-scale stability for about four and a half billion years. The planets whirl around the sun in seemingly unchanging procession, yet we know that the planets and sun have undergone continuous change. For one, the sun spews forth huge amounts of energy daily, reducing its total mass and energy content. The earth intercepts a tiny fraction of the sun's energy, which then heats and helps sustain all life on earth. We maintain our lives by an organization of matter that fixes layers of complexity into a processing scheme that constantly renews us. We take in matter from outside and change its arrangement in order to sustain our own.

Without change we could not detect or know the world; in fact, we could not exist. Our sight depends on light entering our eyes bringing information about the external world. Our touch depends on matter pressing on nerve endings in the skin triggering signals in our brains. Without these varying signals life would be brutal, painful and short. We would drown in every river and fall from every cliff.

If we look up, into our Milky Way and universe, we sense a gentle shower of radiation. Photons from the sun cover us with warmth and life giving energy. The galaxies, invisible in the sun's glare, still touch us with their ancient red-shifted light. From every direction a hundred million neutrinos pierce our body each second; yet we feel nothing and know nothing of their passage. With a singular resolve they pass through space, stars, and men, like ghosts, unscathed.

The Neutrino

Before 1931, neutrinos were unknown. In that year W. Pauli suggested that "a missing" recoil energy from the nucleus of an atom during radioactive beta decay was actually another ejected particle with no charge, little or no mass, and traveling at or near the speed of light. The idea was largely ignored since there was no experimental evidence for such a particle. Then in 1956, Reines and Cowan indirectly detected neutrinos coming from the Savannah River nuclear reactor. More recently (1987), a number of neutrino detectors have been built in deep mines to be free from interfering energy bursts. Even so, in one experiment where hundreds of billions of neutrinos from a distant supernova explosion were expected to pass through a huge detector in a mine, only eleven interactions were recorded. Today we know that neutrinos can pass through the earth more easily than light can pass through a pane of glass. They interact with matter so weakly they are seldom deflected or absorbed. Every cubic yard of the universe contains about a billion. We cannot avoid them; we are being pierced with innumerable neutrinos from the moment of our conception. Nuclear processes in stars emit neutrinos constantly, adding to the huge supply already existent. All are moving relative to every other particle and photon in the universe. Here is change indeed.

The Expanding Universe

To add to this bedlam of motion, the universe itself is expanding. In 1929 the stock market crashed, beginning a terrible worldwide depression. But much more important from the viewpoint of our understanding of the natural world, 1929 saw the publication of experimental data obtained by Edwin Hubble and others showing that the light coming from distant galaxies was redder than light from nearer galaxies. This meant that the more distant the galaxy, the faster it was receding from every other galaxy, including ours. The universe was expanding! Hubble's data had been so carefully obtained and confirmed that scientists were quickly convinced of its truth. Even the great Einstein accepted the expanding universe data and removed the repulsion constant (added to describe a static universe) from his relativity equations.

The earth-based view of the day was that the universe of planets, stars, and galaxies was essentially static. Celestial bodies move in a concerted way, east and west and north and south. Excepting the planets, moon, and sun, the other bodies seem to keep their relative positions. There was, however, another, a third dimension, beginning at the earth and moving out into space. Human vision was not adapted to detect large movements in that direction. Our binocular vision has great survival value when it enables us to decide whether a deer or a grizzly is coming our way, but no value at all in deciding whether the galaxy Andromeda is approaching or receding. Our vision evolved for close in encounters on a three-dimensional Earth. Only when we had spectroscopic telescope observations showing the shift of atomic spectral lines toward the blue or red end of the spectrum were we able to detect movement of the galaxies toward or away from the earth.

The first inkling of wavelength shifts of starlight with changing distance came in 1912 when the astronomer Vesto Slipher found that the atomic spectral lines of light from the Andromeda galaxy were shifted toward the blue end of the spectrum. At the time Slipher did not realize he had stumbled on the Doppler Effect as it relates to light. That effect was first observed with sound waves in 1842: if a source of sound and an observer are moving relative to each other; the sound waves will be compressed if source and observer are moving closer, and stretched if they are moving apart. This has the effect of shortening or lengthening the wavelength of the sound. Many people have noticed that the siren on a fire engine or ambulance sounds high pitched as the vehicle approaches and drops toward the bass as it passes and recedes. It is the relative movement of the source (the siren) and the observer (our ear) at the moment of emission or detection that produces the phenomena. Slipher had found the light wave equivalent of the Doppler Effect. Hydrogen spectral lines from the Andromeda Galaxy were blue-shifted or shortened; therefore, Andromeda was approaching our galaxy.

An expanding universe means that the galaxies are moving away from each other. They are not all moving away from our Milky Way as the center; our galaxy is not at the center. There is no center. The situation is like well mixed raisin bread dough; when baked the bread rises, all the raisins move away from each other. Or it is like a balloon with dots on the surface; if the balloon is blown up, the dots move apart with every puff. Or it is like a rubber string of pearls; when the rubber string is stretched, all the pearls move apart. These are three, two, and one-dimensional models of how galaxies might analogously recede from each other in our four dimensional universe. Unfortunately, it is beyond our experience or ability to form a clear mental image of a four dimensional universe (Hawking 1988, Davies and Gribbin 1992). Since we evolved in a three-dimensional world, our existence depends on our ability to function in three dimensions, not four. With the universal galactic expansion comes another relative movement of every particle and photon in the universe, compounding the other movements we have already discussed.

An expanding universe holds important inferences; the universe is unstable, it is continually changing. Why doesn't the universe hold still; why isn't it static? An explanatory conceptual model is that what we call existence requires matter, mass and energy. For instance, we are made of atoms, which are a form of mass, and energy that interacts with mass. But in that interaction, energy is used up so that a present existence cannot hold indefinitely; a change, a new arrangement must occur. Thermodynamics is the science that traces the role of energy in changing relationships. One of the central discoveries of thermodynamics is that every chemical or physical interaction is accompanied by a change in energy content and increasing entropy. The much quoted Second Law of Thermodynamics states in one of its many formulations that: in an isolated system, the spontaneous flow of heat (energy) is always from the body (mass) of higher temperature to the body of lower temperature. An "isolated system" isolates the mass and energy of the system allowing nothing to enter or leave during the experiment and measurements. The law is statistical; it refers to systems with large numbers of photons (energy units) and particles (mass units). At the end of the chapter is a somewhat more extensive discussion of thermodynamics.

Change in Quantity

Aristotle's second class of change is variation in quantity. Populations, for instance, grow and decline. The population of Ireland and the lower Rhineland declined in the late 1840s due to the failure of the potato crop. European populations fell dramatically in the fifth to ninth centuries during the collapse of the Roman Empire and the early Middle Ages, and again during the time of the Black Death, 1348–50. The number of rings on tree stumps attests to the years of growth. But, although trees grow in height and girth each year, the number of mature trees on earth steadily declines from excessive cutting and careless burning. Erosion and pollution increase. World population increases; some birth rates decline others increase. Human longevity increases, but some populations show a decrease. Overall, human population increases and is on a collision course with the resource limitations of Earth. We are inundated with a mighty flow of numbers, data, and statistics that also grow each year. The sciences prove everything on the basis of the increase or decrease in the numerical value of observable phenomena. We count our health in years and our wealth in coin. Aristotle was right: changes in quantity are a characteristic of things.

Every quantitative change has implications for other events. In AD 9, three Roman legions, the 17th, 18th, and 19th, were trapped and annihilated in a valley of the Totenberg forest. The most efficient and ruthless military machine of the ancient world was stunned and crippled. The three, representing more than ten percent of Emperor Octavius' twenty-seven legions, were often lamented but never reconstituted. The battle of Totenberg Forest is little chronicled in historical records, perhaps because there were few Roman survivors, or because the Romans celebrated their triumphs not their defeats. But the battle represented a high water mark; the Romans never again tried to expand north of the Rhine River. Later the tide turned; Rome was inundated and destroyed by a rush of many peoples from the north and east.

Change in Quality

Change in quality is Aristotle's third category of change. Alterations of appearance, odor, sound, taste, and touch are changes as much as movement and quantity.

The flavor of tomatoes has changed since the 1930s. The pithy, pale and dry three-in-a-box items that show up in the supermarket are of a poorer world. In the 1930s, it was our habit to sneak into grandfather's (or the neighbor's) garden to snatch a full red ripe tomato, rub it free of dried dirt, and take a juicy sweet bite. That was real tomato flavor. If we thought ahead, we brought along a shaker of salt, doubling the pleasure. After about three such gems we were keen for the next adventure. The world seemed brighter, less menacing, and right.

A fine soprano voice is a thing of beauty that announces its own excellence. The connoisseur of wine has a choice of jargon that defines the character of each bottle. Such adjectives as full-bodied, throaty, fine, great, pleasant, sweet, complex, light, foxy, young, dry, flinty, acid and so on, allow wine lovers to express their opinion and delicacy of taste. We all know that heating ice turns it from a hard cold solid into a soft warm liquid and then into a hot colorless gas. Tree leaves turn from green to red, brown, and yellow in the fall. Carbon black is a dull messy material that cannot be wiped away; it is so fine it hides in the very pores of the skin. Graphite, on the other hand, is slippery, glossy, but still messy. Diamond is a clear, brilliant crystal and never messy. These three forms of carbon differ primarily in the way their atoms are arranged.

It is characteristic of changes in quality that a new vocabulary is needed to describe the change. In a chemical reaction it is not enough to say, for instance, that three molecules, two of hydrogen and one of oxygen, have changed into only two of water. Things have disappeared and a new material has appeared. The change in properties and appearance has been dramatic. New descriptive terms must be created to tell us how liquid water differs from the original gases. Changes in quality are also associated with Aristotle's fourth and perhaps most important kind of change, those of coming to be and passing away.

Coming to Be and Passing Away

A beginning and an end to existence is characteristic of all forms of organic life and to inorganic matter as well. The fossil record attests to the numberless individuals who came into existence and passed away and to the species that came to dominance then disappeared. In the inorganic world, the creation of energy and particles such as photons, electrons, and protons in the Big Bang or First Change 13.7 billion years ago; and the production of heavy elements such as iron, carbon, and oxygen in the supernovae of the later universe are examples of coming into existence. The "Big Bang" was a vast coming to be.

Chemical reactions are events that are both coming to be and passing away. Consider the oxygen–hydrogen reaction cited above (at 20 degrees centigrade):

$2H_2 + O_2 \rightarrow 2H_2O$ +68.4 kilocalories/mole

By convention the reactants are on the left and the products on the right side of the equation. Here the diatomic hydrogen and oxygen gaseous molecules pass away and two molecules of water come into being with the evolution of 68.4 Kcal/mole of energy. But what actually passed away; what came into being? The equation balances; the same number and kinds of atoms appear on both sides of the equation. It is the relationship, arrangement, and energy contents of the atoms that changes. The 68.4 kilocalories of energy changed from atomic bonding energy into radiant thermal energy. The atoms of water are in a new relationship which is more stable; therefore, with lower energy content than the former diatomic hydrogen and oxygen gases, which are actually a high energy explosive mixture.

Here, at the atomic level is another world. By arranging and rearranging the 92 naturally occurring elements, the whole vast panorama of our visible universe is constructed. In a like manner we make more than four hundred and fifty thousand English words from the twenty-six letters of the alphabet and from the words we construct phrases, sentences, poems, literature, sciences, and philosophies. And as we shall see later, DNA has only four letters but can form twenty different amino acids, which in turn can fabricate millions of different proteins in our bodies.

In these examples we see that reductionism, which claims to learn much by dissecting things into their constituent parts is at least partly valid. But the extreme reductionist claims that crystals, fleas or sonnets are nothing more than their parts, are wrong. Is the word "true" nothing but the four letters e, r, t, u? A new, stable, and reproducible arrangement of parts brings a new being, a neont, into existence.

By neont (neontic, neonate etc.) we mean a new being as opposed to a reduction into parts or reductionism. If an object can be reduced into constituent parts, the parts must have come together at some moment of coming to be. When the Big Bang occurred, the universe changed from a point of enormous density into hot, undifferentiated, and radiant plasma. Since then, all composite things: atoms, mountains, moles, and movies came into being through processes of accretion and evolution. The wonders of life, family, society, and civilization are the result of an ingathering of parts by processes that produce new entities with new attributes. A large part of any civilization's energy goes into neontic processes whether they create wars, pyramids, schools, Internets, or factories.

Do ideas change? It appears so. Although not material things, ideas are derived from matter in terms of brain cells, words, plans, models, relationships and communication. Consider the idea of atoms. Leucippus of Militus is credited with originating the concept about 500 BC. His more famous disciple, Democritus, developed the idea of irreducible building blocks and named them "atomo" about 430 BC. They were thought to be uniform, hard, small, and indestructible. Without experimental evidence of any kind, the idea was developed from pure reason and a philosophic problem. That problem was, if you take a block of marble and break it in half, then in half again on and on, you still have marble, but finally in powder form. To carry on this process ad infinitum was thought to be impossible by Democritus so he postulated a final division into irreducible "atomo," too small to be seen. In the case of marble he would have ended with atoms of calcium, carbon, and oxygen all of which, we now know, can be smashed into smaller particles such as electrons, protons, and neutrons. But the approach was correct. The greatest ancient Greek philosophers, Aristotle and Plato would have none of it, insisting that matter was continuous, indivisible, and made of four basic elements: earth, air, fire and water.

The argument continued on into the first century BC when the Roman, Lucretius, wrote a long philosophic poem "On the Nature of Things" which described the atomic idea for general readers. But Aristotle's concept of earth, air, fire, and water prevailed through the medieval period. In 1649, Lucretius' poem was rediscovered and widely printed. It was accepted with more enthusiasm since experimental evidence was beginning to give the atom idea a better explanatory role as compared to continuous matter.

Below are a few passages in which Lucretius describes the characteristics of atoms in his poem "De Rerum Natura," first century BC (Humphries 1968):

Book I, Lines 219–221.

> "Our second axiom is this, that nature
> Resolves each object to its basic atoms
> But does not utterly destroy it."

Lines 614–616

> "Something must be the smallest that there is,
> Otherwise, every possible tiny object
> Will be composed of infinite particles."

Book II, Lines 307–308.

> "That while the atoms are in constant motion,
> Their total seems to be at total rest..."

Lines 399–401

"From smooth round atoms come those things which touch
Our senses pleasantly; what feels harsh, or rough,
Is held together by particles more barbed".

These passages tell us that all things are made of atoms, that atoms are the smallest things and in constant motion, round, and smooth or barbed. Of these attributes, only constant motion conforms to modern science.

As more experimental evidence began to accumulate, scientists such as Galileo, Newton, and Boyle lent their support to the atomic idea. In 1794, the chemist J. Proust published his law of proportions, which stated that the atomic components of a chemical compound always combine in definite ratios by weight. In 1808 Dalton published an elaboration of Proust's work that laid the scientific basis of an atomic theory of chemistry. Aristotle's four elements slowly faded away and the nineteenth century saw the steady advance of atomic theory. But little was actually known about atoms; they were still small, hard, round balls.

Based on alpha particle scattering experiments by Geiger and Marsden in 1911, Rutherford proposed that atoms consisted of a small central nucleus that contained a positive charge and balancing negative electrons circling the nucleus. Bohr used this nuclear model to develop his theory of the hydrogen atom in 1913. Progress was accelerating; the shell model of atoms was developed. Now the wavelength of atomic spectral lines could be calculated and were shown to be the same as observed values. Quantum models of the atom with greater predictive and explanatory power quickly followed.

Any history of an idea will show incremental changes similar to those of the atomic concept, briefly recounted above. The idea of evolutionary change as the process by which all creatures had reached their present state was spreading decades before Darwin published his epochal work "On the Origin of Species By Means of Natural Selection" in 1859. Cuvier, a French zoologist, established the science of anatomy and assembled a number of fossil skeletons that showed that many species had become extinct. By about 1800, he had ascertained the anatomical relations between several different species. He also described the relationship between the depth of rock strata and the similarity of the contained fossils to modern species. The greater the rock depth, (i.e. the older), the less similar are the fossils to modern species.

Lamarck, another French biologist, published a theory in 1809 proposing that acquired traits are inherited by individuals and that organs are strengthened by use and weakened by disuse. The process is usually described by the example of the giraffe's long neck. Giraffes browse on tree leaves; the individuals with the longest necks have improved chances of survival since they can reach higher. Constant stretching of the neck causes it to get even longer and stronger. The acquired longer neck is then inherited by the giraffe's offspring and the cycle is repeated. This was the first theory of how a natural process could bring about complex anatomical changes. Darwin's own grandfather, Erasmus Darwin, wrote a long poem in 1796, the Zoonomia, which discussed evolution in a vein similar to Lamarck's. After Darwin published the theory of variations and natural selection, Lamarck's theory of acquired traits slowly fell into disfavor, although its shadow has been revived several times in the last 150 years.

The study of history is largely a recounting of how mankind's changing beliefs, religions, philosophies, myths, laws, traditions, methods, and practices influence a society's organization and function. There is a kind of contagion, which can operate in any direction; a new idea or process can spread from an individual to the ruling elite, to the general population, an isolated group, or the reverse.

As an example of changing ideas spreading from an isolated group, consider the diffusion of Christianity into the mainstream of Roman civilization from obscure origins in the Middle East. The Christian belief spread slowly after the crucifixion of Christ in AD 30. Christianity was originally in competition with several other groups such as Gnosticism and Mithraism. But Christianity won ascendancy over these two so-called mystery religions because of better organization and the favor of Constantinus, a member of the Roman military elite.

Constantine was to become the emperor Constantine I. He and Licinius, emperor of the eastern half of the Roman Empire, ended persecution of the Christians by the Edict of Milan in AD 313. After a falling out and the defeat of Licinius in 324, Constantine attributed his victory to the Christian cross which some of his legionnaires had painted on their shields. After conversion to Christianity and as sole emperor, Constantine called the First Council of Nicaea in 325 to codify the beliefs that define Christianity. This effort resulted in the Nicene Creed still in use among Christians. He was active in the propagation of the faith until the time of his death in 337 and must be given the greatest credit for the ascendancy of the Christian ideal. As the dominant member of the ruling class, he made conversion to Christianity acceptable and advantageous to all.

The First Crusade shows how an idea arising mainly from the ruling clergy and aristocracy can spread into the general population. Europe by 1095 was in a condition of prosperous ferment. The feudal states were constantly at war to the dismay of the pope and the detriment of commerce. The region was awash with unemployed knights looking for land to conquer and hostages to take. When the Byzantine Emperor Alexius I turned to the Roman pope, Urban II, for help against the invading Seljuk Turks, it seemed like an opportunity to solve several problems. At the Council of Clermont in 1095, which approved a canon granting indulgence to those who would fight the infidel Turks, Urban II called for a crusade to free the Holy Sepulcher from the Muslim Turks. His speech was given to a large outdoor crowd. Although there is no written record of the speech, it was said to include a call for knights to stop their feudal warfare and put their skills to a greater purpose. With Urban II nurturing a vision of a free Holy Land, the response was immediate and widespread. Armed bands of men, at first from France then spreading to other European areas, gathered and began a disorganized but relentless march toward the Middle East. Thus began the two hundred year bloodbaths known as The Crusades. After eight crusades and many smaller incursions, with slaughter and massacre on all sides, the seesaw struggle was ended when the Crusaders were expelled from the Holy Land in 1291 after the fall of Acre, their last fortress city.

The pervasive nature of change could be endlessly extended. All history is a recitation of changing events. The sciences are systematic investigations of changing physical occurrences. Music, speech and the written word are ever changing streams of sound and thought. But, are there things that never change, perhaps the atomic notions of spin, charge, or mass or Plato and Berkeley's eternal ideas and essences or philosophic absolutes or the laws of conservation of mass, energy and momentum? The answer is no; with the "Big Bang" there was a change from nonexistence to existence, to one of coming to be. Everything we observe came into at least a seminal existence at that moment. Whether there will be a "Big Crunch," or a relentless dilution of matter into a featureless haze, or a conversion into a static universe in the future, we do not know. But recent data favors an accelerating expansion as the most likely and the stability of the basic constants of nature have also recently come into question.

Change and Duration

Why is it that we hear, see, and feel no change as we sit here reading and writing? Perhaps we can feel our hearts beating and, if we listen closely, hear our breathing; but what of this table we are using? It does not change; it only sits there; it endures.

We know that at the atomic level, atoms and molecules are all vibrating. But at the macro level, where our senses function, there is no sign of such activity. The table is at a different more enduring human level than atoms. If we consider the attributes of a table only as a table, do they change? A table is atoms, molecules, and wooden boards; it is also a shape or form. Plato used this feature of things to postulate his theory of Ideal Essences or Eternal forms that never change. Plato's Eternal Ideal Essences were one of the first injections of absolutes into the rational thought of Western civilization with consequences that are still with us today.

When we look out into the world, things do not seem to change that much. The oak tree outside the door changes with the seasons, but we recognize it each day as the tree we knew weeks, months, and even years ago. Different things have different rates of change. We measure such rates of change on the basis of how often we observe distinctions. It is a comparative method; how long did the daffodils bloom compared to the peonies? Things have this property of relative endurance or duration compared to the rate of other events and happenings. Since there is a vast overlap of individual durations, however, we tend to see the world as largely continuous and static.

Duration is often characteristic of classes of things. Buildings last for decades or centuries; azaleas bloom for weeks. Each depends on stability within a given environment. The azalea blossoms depend on a steady flow of water, nutrients, and moderate temperatures to maintain themselves. Buildings depend on an unvarying physical situation. There should be no earthquakes, leaks, breaks, settling cracks, fires, or vandals.

Enduring inanimate things are in a state of low energy. By definition, low energy means they are stable within their present environment. Large amounts of energy must be used to change them. Atoms and molecules are in a low-energy, stable state if not many other atoms or molecules will react with them. For instance, silica, which is widely distributed as the enduring sand and rock, can be quickly dissolved only by the strongest acid, hydrofluoric, or by fusing with alkali.

The Source of Change — Thermodynamics

The ancients largely accepted the notion that change was universal, although the Greek philosopher Zeno disagreed. Over the last one hundred and fifty years science has developed theories that explain change as due to the interaction of mass and energy. Mass we know well enough. If we didn't have it, we would float off into space. Every bit of mass attracts every other bit of mass through the force of gravity. Energy is another matter; energy takes many forms that tell us little about what they really are. What kind of mental images arise when we think of energy such as, kinetic, heat, chemical, radiant, atomic, gravitational, electrical, or potential energy? Although each form of energy has been studied and has a very exact mathematical formulation, they are often abstractions that evade our earth-evolved brains and reasoning powers.

Kinetic, is the energy of matter in motion; chemical is the energy evolved or absorbed in a chemical reaction between atoms; potential energy is a property of a body due to its position. An apple at the top of a tree has greater potential energy than one at the base of the tree. Recall that Newton reputedly discovered the gravitational energy concept when he observed an apple drop to the ground. When it was shown during the eighteenth and nineteenth centuries that all forms of energy could be transformed into heat energy, methods were sought to relate heat energy from the different sources. This resulted in the development of the principles of Thermodynamics, which refers to the movement of heat energy from one body to another. When you drop an ice cube into a cup of tea to cool it, heat energy flows from the hot tea into the ice cube and melts it. Heat energy flowing in a metal bar from the hot end to the colder end means that the large atomic vibrations (thermal energy) at the hot end decrease as energy flows to the cold end until temperature equilibrium along the whole bar is reached. This is called heat energy conduction, one of the three forms of heat transfer; the other two are radiation and convection.

Radiant heat energy transfer occurs when we warm our hands before an open fire or stand out in the sunlight. Convection occurs when one material flows past another. For instance, a hot gas flowing through a cold metal pipe will quickly heat the pipe as the energy vibrations from the gas pass to the metal atoms in the pipe causing them to vibrate more rapidly thereby lowering the energy content and temperature of the gas.

Interactions of mass and energy accompany all changes in the universe. These universal changes are described by phenomena that are related in the Laws of Thermodynamics. The First Law, which is also the Law of Conservation of Energy, states that energy can be neither created nor destroyed. Since energy (E) can be converted into mass (M) and vice versa according to Einstein's famous equation $E=MC^2$ the law also applies to mass (M).

The second Law of Thermodynamics states that, in an isolated system, heat energy cannot flow spontaneously from a cold body to a hot body or conversely heat only flows from a hot body to a cooler body. There are many formulations of the two laws; Feynman (1963) prefers:

First Law — The energy of the universe is always constant.

Second Law — The entropy of the universe is always increasing.

Gribbin (1996) jokingly summarizes everything into: 1. You can't win; 2. You can't even break even; and 3. You can't get out of the game.

Both Einstein and Eddington have stated unequivocally that the laws of thermodynamics are the most basic laws of physical science; consequently we must accord them our full attention. The concepts involved are both abstract and complex but it is worth taking the time to create mental images or models that allow us to better understand the phenomena.

First, the two laws are empirical laws; they are based on data and facts, which grew out of experiments performed originally in the eighteen hundreds. No observations have ever been found that contradict the laws. But the experiments must be carried out with extreme attention to experimental detail. The laws have been formulated in a number of variations; a few of them follow.

First Law of Thermodynamics

The heat energy (Q) put into a closed system plus the work (W) done on the system, is the total increase in internal energy (E) of the system.

$$Q + W = E$$

A closed system is any group of bodies which are separate from and cannot absorb energy from the surroundings. The concept of work (W) in this usage is the force (F) necessary to move a mass a definite distance (D), or algebraically:

$$W = F \times D$$

When you work to dig a hole in the ground, you are using force to raise a mass of dirt a certain distance up and out of the hole against the force of gravity.

Second Law of Thermodynamics

The entropy of the universe is always increasing.

Or, the spontaneous flow of heat energy in an isolated system is always unidirectional from the higher to the lower temperature body until temperature equilibrium is reached.

Or, in an isolated system, it is impossible, on balance, to convey heat energy from a body at one temperature to another body at a higher temperature.

Or, spontaneous processes occur only with an increase in entropy or, the direction of spontaneous change leads to an entropy increase.

These definitions of the second law apply to statistical systems containing large numbers of particles like atoms or photons, not to individual particles.

Entropy

Entropy, the major concept of the second law, is often equated to order and disorder or dispersal of energy, or degree of randomness in a system. Entropy is defined mathematically as increasing when disorder increases and decreasing when order increases. Mixing two pure materials together, for instance, causes entropy to increase because a single pure material is in a more ordered state and less dispersed than a mixture. Increasing entropy has also been described as increasing dispersal, disintegration, dilution, and a decrease in the ability of a system to do work.

Since the entropy of the universe is constantly increasing we have another example of universal change. Increasing entropy also degrades the ability of a system to do work; it is a measure of exhaustion and impotence, we should note; however, that some systems such as living creatures bring about decreases in entropy, becoming more ordered and less dispersed. This comes about because living creatures consume high energy food and discard low energy waste, thereby increasing the entropy of their surroundings while decreasing the entropy of their bodies.

Summary

And finally, now, in this seemingly stable epoch, we find ourselves walking on the surface of a large oblate sphere, which is spinning on its axis while rotating around a sun. The sun is half way along a large spiral arm of our galaxy, the Milky Way, which is also rotating once every 200,000 years and moving away from all other galaxies (excepting a few like Andromeda). The intervening space between galaxies is filled with neutrinos, photons, particles, and dust, all moving in their appointed ways. What is at rest? What is unchanging? There is nothing at rest or unchanging, at least not in this universe.

Bibliography

Aristotle (384–322 BC), *The Great Books 1952, Physics*

Davies, P. and Gribbins, J. (1992), *The Matter Myth*

Encyclopedia Britannica 15TH ed. (1986)

Feynman, R., 1963, *Lectures on Physics*

Gribbins, J., 1996, *Companion to the Cosmos*

Hawking, S. W., 1988, *A Brief History of Time*

Humphries, R. 1968, *Lucretius; The Way Things Are*

Chapter 8. Scientific Reality

There is nothing mysterious or magical about how science uncovers nature's various levels of reality. The same five senses and powers of reasoning we use in normal life are applied but with more diligence, analysis, and precision. Each expansion of knowledge nourishes further expansion. The technological development of new materials and processes encourages the fabrication of improved scientific apparatus. These in turn make more sophisticated experiments possible. It is an accumulative process; new ideas, theories, and experimental techniques lead on to even newer ideas, theories and experiments — things change.

Dual Reality and Reductionism

We should understand that, for us, there are actually two universes, the one we are made of — the physical universe, and the one we live in — the biological universe. The physical universe is composed of mass and energy and operates with no known purpose. It endlessly cycles through repetitive interactions with mathematical precision yet with increasing entropy content and disorder.

At the present time, the facts uncovered by physical science (Gribbin, 1996; Kaufmann, 1985, Greene, 2004) present us with a broad conceptual model of a physical universe that began about 13.7 billion years ago as a point of extreme density. This point, or singularity as it is called, represents a time when the present laws of physics did not apply and time and space did not exist. Note that a point of extreme density implies a point or region equivalent to all the mass and energy we now find in the universe. With this "Big Bang" or first event, the theory says, the singularity suddenly began expanding as a fiery region of high temperature matter, the quark-gluon plasma. During the first tiny fraction of a second the plasma grew in an inflationary burst from less than the size of a proton to about that of a grapefruit. Inflationary theory has been helpful in answering many questions surrounding the quantum and cosmological nature of the early universe. The source of the singularity, if it had a source, is as yet mysterious, unexplained, and unknown but it was extremely concentrated, ordered, and of low entropy.

As the universe continued to expand, the temperature dropped and electromagnetic energy waves began to change into simple forms of mass such as electrons, positrons, and neutrinos with a few protons and neutrons. Later these began to interact and form simple atoms such as hydrogen and helium. Since all mass attracts all other mass through the universal property of gravitation, particulate matter began to condense into increasingly large bodies, eventually forming galaxies, stars, planets and all the other material features of the universe.

The universe of life reveals a very different reality. It began almost 4 billion years ago on the planet Earth and is also composed of particles (atoms arranged in large and complex molecular arrays). In addition, it has a goal and a purpose. Its purpose is to continually maintain and reproduce itself from the raw materials of its surroundings. The reproduced individuals, however, due to errors and defects, are not exactly like the parents and differ in physically possible but chance ways. These changed individuals, as they undergo the natural selection process, are "selected" if they can meet the challenges of their environment at the time; those that cannot, die. Living creatures, in one word, evolve.

Between physical mass and energy and biological individuals there is an immense chasm. Although all living creatures are made of atoms that obey physical laws such as gravity and thermodynamic laws, they possess a host of traits that are not exhibited by the physical universe anywhere or in any way. We could mention love, joy, pain, inheritance, selective interaction with the environment, the will to live, fear, dread, and so on.

There has been a reductionism argument among scientists and their critics that the characteristics of a complex system such as a tree or human can be fully explained by knowledge of their basic or smallest constituents (Mayr, 2001; Adler, 1985; E.B. 25, 1989). This is the critic's charge that scientists claim trees and humans are nothing but atoms. Such critics abhor the apparent bleakness of scientific concepts on ideological grounds; that is, they don't like them. They don't question the truth of the facts; rather they fear the possible deductions and inferences from the facts. These reductionist charges are false because of two phrases, "fully explained" and "nothing but." These phrases are exclusionary, absolute, and contradict the known nature of life.

Yes, humans are made of atoms but are somewhat more than that; have you ever seen atoms hold hands and go for a Sunday walk or listen to Mozart? Do rocks, or stars, or galaxies? No, we are each unique among the stars; never seen before, never to be seen again. There is no knowledge that fully explains us. Life forms are the perfect example of the principle that "the whole is greater than the sum of the parts." They exhibit qualities that none of the parts have and have few of the qualities of parts. But we also cannot measure or fathom in any way how much "greater" or "better" living creatures are than their parts. Parts and wholes are incommensurate.

Some elementary particle physicists on the other hand, embrace reductionism because science is a hierarchy and the elementary particle is its foundation; it is therefore, most fundamental, important, and worthy of support by the government. An example is the Superconducting Super Collider favored by some (Weinberg, 1992). That may be true for the physical universe but it is not for the biological universe.

The most important foundations of biology are evolutionary theory, natural selection and the structure and function of DNA, deoxyribonucleic acid. Detailed knowledge of DNA's attributes will revolutionize life, particularly human life, by correcting genetic errors, and tendencies toward degenerative diseases. But this will happen only if we embrace DNA research with enthusiasm and an open mind toward its human applications. For humans, it matters little whether we are made of atoms or apple seeds; we suffer the gamut of human conditions and emotion; we are still hungry, sad, happy, suffering, fearful, or amused, yet feel no atomic malaise.

Science and Numbers

Science is largely a numbers game: how far, how fast, how much, acceleration, pressure, voltage, energy, time, and how many, dyne-centimeters, degrees, lumens, and so forth. In fact, science has little to say about the physical universe that is not contained and explained with numbers. The attraction of numbers for science is that they lend themselves to brevity in the manipulation of mathematical and deductive relationships. Mathematical relationships lend precision, repeatability and proof to scientific ideas.

As science advances, scientists make more use of mathematical notation and specialized jargon to shorten and simplify their written reports and discussions. Words and contractions like spectra, entropy, enthalpy, racemic mixtures, isomers, Rydberg constant, dyne-cm, joules, supernova, quarks, EPR, EMF, DNA, GMW, and half-life have become the norm. These represent numbers and concepts that grew out of extensive experimental trials, but they have caused an increasing poverty of understanding between scientists and laypersons. Even discussions between scientists become opaque with increasing use of jargon.

The role of numbers and mathematics in science led the physicist Eugene Wigner to publish a paper in 1960 entitled "The Unreasonable Effectiveness of Mathematics in the Natural Sciences" (Davis and Hersh, 1981). Wigner's question was: how can nature's phenomena be reduced to numbers, and then incorporated into mathematical relationships that give correct predictions? One of the responses to Wigner's question is that mathematics is a creation of the human mind comparable to art, philosophy, Laissez faire capitalism and so forth (Hersh, 1997). A second response which captured the mathematical mind early and is still the predominant view today is Plato's essences or ideals. The ancient Greeks Pythagoras (540 BC) and Plato (380 BC) held that numbers and mathematics had an ideal, nonmaterial reality that existed before and outside of space and time. The idea of beauty, for instance, materializes when it comes into contact with matter. A real cow is only a projection of the mass less, eternal, ideal cow. Here is ultimate absolutism in full flower.

The basic method of mathematics is described by Hersh (1997) as "conjecture and proof." This is the method of science and trial and error in mathematical terms. The mathematician tries to guess intuitively or imagine a solution to a question or problem and then shows how the solution can be used to provide an answer to the question. If consistent with, and not contradictory to the problem or question, the solution is proved. This procedure is like the scientist's hypothesis that is tested against the observable universe for confirmation.

Scientific Models

Each scientific word has a carefully defined meaning that refers to specific features of the natural world. Scientists often create models or mental constructs that allow them to talk and think about phenomena in terms compatible with our three-dimensional, earth-evolved intelligence. Previously we mentioned one, two and three dimensional models that are analogous to a four dimensional universe. Our earth-evolved brains cannot directly visualize four dimensions, probably because there was never an evolutionary need for four. We conceptualize natural phenomena in imaginary models that can act as bridges to our inner feeling of understanding. But we must never forget they are human models only and do not in any sense correspond exactly to the real world. Conceptual models are often stepping stones to more accurate and useful models that further increase our understanding. Models tell something about what we observe and measure but often more about ourselves than about reality.

THE ATOMIC MODEL

One of the most useful scientific models underlies modern atomic theory. J. J. Thomson took the first large step forward while experimenting with gases in evacuated glass tubes having metal electrodes on each end. When an electric potential was applied across the tube with the pressure reduced to about .0001 mm of mercury, there was a glow at the cathode (the negative electrode). The glow was called the cathode ray. These rays are produced by residual gas ions that are attracted to and strike the cathode with considerable force. When the cathode rays were passed between electrically charged plates, they were attracted to the positive plate and repelled by the negative. Thomson deduced the rays must have a negative charge (opposites attract, like charges repel). The rays (actually electrons) were produced by a wide variety of electrode materials and residual gases so he also deduced that the rays were part of all atoms. Finally, if electrons come from atoms and atoms have no charge, there must be a positive charge somewhere in the atom that balances and neutralizes the negative charge of the electrons.

All of this was a multiple deduction. What did it mean? Thomson hypothesized that the positive charge was like a mush or gas in which the electrons were individually embedded. Thomson's hypothesis (1897) became known as the plum pudding model of the atom. About 1911, one of Thomson's former students, Ernest Rutherford, fired a stream of positively charged alpha particles through a gold foil to determine what was inside the mush. Using Thomson's model he thought the positive alpha particles would go through the positive mush with ease since the two would repel each other. Most of them did, but some were deflected away from their original path and a few bounced backward. Rutherford deduced from these results that the atom was mostly empty space with a tiny positively charged central core with counterbalancing electrons orbiting around it. This was the solar system model of the atom.

Neither the pudding nor solar models can explain how and why light with specific wavelengths is radiated from atoms that are excited by heating. It was Niels Bohr, formerly a Rutherford student, who postulated that the electrons in an atom are restricted to specific "allowed" orbits or energy states as they whirl about the nucleus. When an atom expands to an excited state by an electric arc for instance, an electron can be raised to a higher energy level (i.e. farther away from the positive nucleus). Since the high-energy state is unstable, the electron soon falls back into its lower equilibrium energy level and radiates a photon (a quantum of light energy). The radiated photon satisfies the requirements of the First Law of Thermodynamics that energy can be neither created nor destroyed. The color (or frequency and wavelength) of the photon depends on the energy difference between the high and low orbits or energy states.

Bohr's mechanism (formulated in 1913) worked perfectly for hydrogen, which has only one proton in the nucleus and one orbiting electron. His calculated light spectra (wavelengths and frequencies) duplicated exactly the observed spectra of hydrogen. However, for more complex atoms like iron or oxygen, the theory failed. Over the course of the next several decades, Bohr, Pauli and many others modified and extended the allowed electron orbit concept to all elements of the periodic table known at the time. But in order to extend Bohr's concept, the orbit idea was slowly discarded for a "probable" location for each electron. The electrons were visualized as having exactly defined quantum energy states but no exact positions. The quantum idea was strange and disconcerting to most scientists. How can you know you have an electron and not know where it is? The answer is that you don't know. Things get foggy at this level. You can determine mathematically the probability of an electron being in a particular place with a certain velocity but not at a fixed position or velocity. For one position there might be a 20 percent probability of an electron presence, at another 75 percent.

The result of using probabilities led to the quantum mechanical model of the atom. This places protons and neutrons in a central nucleus and electrons far outside of the nucleus with definite allowed energy states but no definite positions or orbits. The use of quantum mechanical concepts for atoms fostered their use in other scientific areas involving subatomic particles. Scientists were able to overcome their aversion to "probabilities" and found the concept to have many advantages. This science of the very small evolved into the most dominant and successful theory of the 20th century. Many remarkable quantum applications involving particles, atoms, and photons were discovered. For a human evolving on earth and made of atoms the quantum atom is of overriding importance in a general way.

The Quantum Paradox

Although unrecognized at the time, the first glimmer of a quantum future came with conceptual differences between Christian Huygens (1690) and Isaac Newton (1704). Huygens argued that light has a wave structure (see Figure 1). If it is a wave, light should exhibit the phenomena, common to water and sound waves, of diffraction and interference. When water waves hit and bend around a pier piling or other obstacle, they are exhibiting the property of wave diffraction. If light waves did this, shadows should have bright and dark banded edges. Interference occurs when two stones, for instance, are dropped side by side into a still pond. Where the resulting water waves intersect, they interfere to form new and different waves. If two wave crests coincide, the amplitude of the resultant wave is increased (constructive interference) and it is decreased if the crest of one falls on the trough of the other (destructive interference). But Huygens' light waves were not known to exhibit diffraction or interference phenomena; therefore, it was obvious they could not be waves.

By contrast, Newton, who had written a major treatise on light (Optics 1704), argued that light was made of particles. He pointed out that if you look behind an object in bright sunlight, you see a sharply defined shadow corresponding to the shape of the object. Where there is a bump on the object, there is a bump on the shadow. There is no trace of the light and dark interference bands expected if light had a wave structure. With light made of particles you would expect to see sharp shadows. The two different concepts held by Huygens and Newton were the beginning of the wave/particle paradox that was to increasingly plague physics for the next three centuries.

Since no one had found interference fringes associated with light waves, Newton seemed to have the better argument at the moment. It was almost a century before Thomas Young, an English physician and physicist found the missing interference fringes and explained them correctly in 1801. Young admitted a fine beam of sunlight onto two closely spaced pinholes in a baffle. The light passed through the pinholes into a dark room where it formed two light cones that fell on a screen. Where the images overlapped, the missing light and dark interference fringes were observed (Wood, 1934; E.B. 1989). Huygens' wave theory of light was finally vindicated. For diffraction and interference phenomena to appear, the wavelength of the waves must be near the size and spacing between the obstacles — in this case the pinholes. That requirement was not recognized in Newton's time.

G. B. Airy extended Young's findings in 1834 by sending light through a single iris diaphragm toward a screen while decreasing the diameter of the iris. At first there was a uniform bright spot on the screen but as the opening grew smaller a dark spot surrounded by concentric dark and light rings appeared. At even smaller diameters, the pattern became increasingly faint and again uniform. Airy analyzed his results mathematically as due to a combination of wave diffraction and interference.

Few of us have an adjustable iris but we can all confirm Airy's observations by making a fine slit about two inches long in a piece of aluminum foil or opaque paper. If we look through the slit held about two inches from an eye at a bright light while focusing on the slit, we will see a series of bright and dark lines parallel to the slit. These are equivalent to Airy's rings and directly confirm the diffraction and interference of light waves. Even more simply, if we look at the space between two fingers held in front of a bright light, we see similar light and dark bands running parallel to the slit between the fingers as we press them closer together (Gribbin, 1995).

Even more support for the wave theory appeared when J. C. Maxwell proposed in 1873 that light phenomena could be completely described as electromagnetic waves. He included visible light, radio waves, gamma rays, x-rays, etc. as electromagnetic waves. The theory was set out in a few elegant equations that included the speed of light as a constant (186,000 miles per second). Einstein was later to use the absolute speed of light regardless of the speed of the observer, as an important part of his relativity theories.

We learn from Maxwell that light has an electromagnetic wave structure. It is a sine wave with periodic oscillations of electric and magnetic components that are at right angles to each other and to the direction of travel. Sine waves are often used to represent natural cyclic phenomena. Figure 1 is a two dimensional representation of some aspects of a single three-dimensional electromagnetic sine wave or photon. The (x, y) coordinate arrows show the varying amplitude or strength of the electric and magnetic fields and the arrow (z) the overall direction of wave motion. Beginning at the left O, the electric field strength is at zero energy; as it moves toward the right it increases in energy to a maximum (A) then decreases to zero again and then to negative values (B) before returning to the zero line at (C). The total curve represents a single wave of the electric field portion of an electromagnetic wave train. The magnetic component of the light wave follows a similar cycle but at right angles to the electric component.

We also know from Planck and Bohr's theory that light has a particulate structure. When Bohr's electrons fall from a high energy excited state to a lower energy state in an atom, a photon is radiated that has an energy corresponding to the difference between the two energy states. These energy bundles or quanta have a beginning and end and behave like particles under certain circumstances. Huygens and Newton were both correct. Light has a wave or particulate nature depending on how you measure them, that is, what kind of experimental equipment you use.

THE PHOTOELECTRIC EFFECT

Toward the end of the nineteenth century some scientists thought science was complete; all we had to do was to carry our measurements of the Newtonian universe out to the fourth decimal place. But an avalanche of new observational data was on the horizon that would require an immense effort of imagination and analysis. After all it was not really clear at that time what electrons, photons, or even atoms actually were.

We will arbitrarily choose Phillip Lenard's discovery of the photoelectric effect in 1899 as the beginning of a new vision. Leonard built on some earlier observations by Heinrich Hertz in 1887 concerning electrodes irradiated with ultra violet light. Lenard observed that electrons, then known as cathode rays, were expelled from some elemental metal surfaces when irradiated with light. The number of electrons expelled was proportional to the intensity of the light. Later, in 1902, Lenard found that the maximum kinetic energy of the expelled electrons depended on the frequency of the irradiating light (photons with higher frequencies have higher energies).

At about this time (1900), Max Planck proposed that Maxwell's electromagnetic fields were not continuous as expected but made of energy packets or quanta. Planck was studying radiation from a black body, which is a body that totally absorbs and reemits radiation thus producing an equilibrium condition at any one temperature. Drilling a small hole into a dense block of carbon can make a workable black body. Light radiating from the hole has a frequency (or wavelength) distribution characterized as black body radiation. When Planck tried to plot the frequency distribution of the radiation coming from the hole observed at different temperatures, he had to treat the data as if radiation existed in discrete energy packets. Without this "quantum" idea, the data made no sense. The then current concept of continuously variable spectral frequencies radiating from a black body did not agree with the facts observed by Planck.

In 1905, along with his papers on relativity and Brownian motion, Einstein also wrote a paper on the photoelectric effect. He argued that Lenard's two laws concerning the effect of varying light intensity and frequency on electrons in metals could be combined with Planck's energy quanta to explain the photoelectric effect perfectly. He proposed that when a quantum of light strikes an electron of an atom in the metal surface, it could be absorbed. With this additional quantum of energy, the electron is able to break its bonds with other atoms in the surface of the metal. Photons with higher frequencies eject electrons with greater energy and photons with higher intensities of a single frequency eject comparatively more electrons. The observed facts agreed with the theory. Or should we say the theory agreed with the facts?

The wave theory of light was in the ascendancy during much of the nineteenth century. However, a major cloud appeared with the Michelson-Morley ether drift experiments of 1887. Scientists assumed that a light wave must have a medium to carry it, as do other waves. Air molecules carry sound waves; when suspended in a vacuum you can hear nothing. Water molecules moving up and down, create and sustain waves as they move along the surface of the water. The general opinion was that an ether with very specific properties filled space and was the medium that carried light. Presumably as the earth spins it must drag the ether with it. This would change the measured speed of light along the equator as compared to a north/south measurement. Using the extreme sensitivity of Michelson's newly developed light interferometer, Morley and he found that whether you measured parallel to the equator or to the north/south axis, you always got the same result — light travels at 186,000 miles per second. The earth did not drag the ether along because the ether did not exist. Light needs no ether; it can travel in a complete vacuum.

Going back to Bohr's atomic model of 1913, the quantum now entered an intuitive and sometimes ad hoc period of extension and formulation of atomic models based largely on atomic spectra. The chaos of the First World War put an end to further advances in theory. The first flowering of the quantum concept was over.

Quantum Waves

Lois de Broglie, who had served in the communications arm of the French army and was conversant with electromagnetic theory, began the second phase. In 1923 de Broglie published a paper suggesting that if light waves can act as particles, then perhaps real particles, such as electrons and protons, can act as waves. This notion of "matter waves" led directly to Erwin Schrödinger's theory of wave mechanics in 1926. Almost immediately (1927) Davisson and Germer experimentally observed that electrons can be diffracted by crystals of nickel and show interference effects. Since diffraction is a property of waves, electrons and other subatomic particles must also have wave-like characteristics as de Broglie predicted.

The decade of the twenties was a period of unparalleled creative activity in quantum theory. In 1925 Werner Heisenberg published a paper on matrix mechanics that showed how to calculate the observed frequencies and intensities of atomic spectral lines. In 1926, Schrödinger introduced wave mechanics and the "wave function" to describe the interactions of particles and waves. But the wave function had no observable attributes and was therefore of limited use. However, within a year Max Born showed that the value of the wave function squared could be used to calculate the "probable" position of a particle within a given region. Probability was not an attractive idea to most scientists, who like things to be precise and exact, but over the years they have learned to live and prosper with it.

P.A.M. Dirac developed a third theory, called the transformation theory of quantum mechanics, while still a graduate student. It was a general and logically simple theory. Within a year three very powerful quantum theories had been developed. They were closely related even though they used very different mathematical formulations.

Heisenberg published his Uncertainty Principle in 1927. This states that both the position and momentum (mass times velocity) of a particle cannot be known precisely at the same time. Formulated mathematically, it states that the uncertainty in measuring the position of a quantum particle (Ux) multiplied by the uncertainty in measuring its momentum (Umv) must be equal to or greater than Planck's constant (an extremely small energy constant equal to 6.55×10^{-27} erg-seconds). Heisenberg's principle has been confirmed by innumerable experiments. All scientific measurements are subject to uncertainty and error but good procedure requires the experimenter to determine the amount of error, usually by repeating the experiment many times. But quantum experiments are different; they are limited in accuracy regardless of the quality of the experimental apparatus or the care of the experimenter. There are some things we can know only in a rather fuzzy way.

Meanwhile, in Copenhagen, Niels Bohr had been working on atomic and quantum theory since before 1913. Over a number of years scientists concerned with the interpretation of quantum theory and particularly the growing indeterminacy of the theory had gathered around him. With Bohr, scientists such as Dirac, Jordan, Born, Heisenberg and Pauli developed an assessment of quantum theory that became known as the Copenhagen Interpretation or CI. Their interpretation was based on the indeterminacy principle and resulted in the concept of complementarity. Bohr became the spokesman for CI against all detractors.

Complementarity was CI's way of reconciling the particle aspects of photon waves and the wave aspects of particles. Bohr argued that wave and particle complement each other and are enclosed in Schrödinger's wave function in a kind of undefined limbo until a measurement is made; then the wave function "collapses" and a particle or wave appears depending on the experimental procedure. According to classical thinking, a particle is an entity confined within an extremely small volume;

examples are electrons, protons and neutrons. A wave was thought of as a phenomenon that spreads out to fill all space like ripples spreading from a stone dropped into a still pond. Classically, a particle is a particle and a wave is a wave and that is that; but quantum particles and waves are different.

The Flowering of the Two-Slit Experiment

During this period of rapid theoretical development, there was a parallel development of experimental techniques aimed at a more complete uncovering of the quantum state. Some of the improvement in experimental technique came from the substitution of two slits for the earlier two pinholes. What scientists found only deepened the quantum puzzle.

Before we go into the details of the two-slit process, we must clarify the nature of diffraction and interference of light waves somewhat. Light is diffracted by all sharp edges; a two-slit apparatus is not required. The effect is so small we cannot observe it with the unaided eye. But in the controlled confines of the laboratory, diffraction can be seen clearly, particularly if collimated or monochromatic light is used. For instance, monochromatic light shining on a razor blade will show a halo around all the razor edges. These are diffraction lines. They are normally too faint and too small for our detection by eyesight alone.

When a two slit apparatus is used to observe light waves or atomic particles, the wave fronts spreading out from the two slits interfere to produce prominent diffraction patterns of light and dark bands. A schematic drawing of a two slit apparatus is shown in figure 2. The lines A–A and B–B represent baffles that are at right angles to the plane of the drawing. Light rays moving left to right first impinge on S1 (a slit in A–A perpendicular to the page), which serves to collimate the light into a narrow beam. The curved line to the right of A–A represents the spreading wave front coming from the slit S1. The wave front spreads onto the baffle B–B and two new wave fronts spread out from the slits S2 and S3. Where these wave fronts overlap to the right of B–B, they interfere destructively or constructively depending on whether the wave amplitudes cancel or reinforce each other. As a result, the interfering waves exhibit light and dark intensity bands represented by the complex curve to the left of the screen C–C. The two curves to the left of D–D show the intensity of light received at the screen when only one of the slits S2 or S3 is open. These curves can be permanently recorded on a photographic plate or optical scanner and can be detected by eyesight.

When photons are sent through only one slit, S2 for instance, while S3 was still open, human reason and logic abruptly collapsed. The complex diffraction curve of C–C appeared again! How did the photon stream passing through S2 sense that S3 was open? And even if it "knew," how or why did it form a diffraction pattern like C–C? What were the S2 photons interfering with? If we add the two curves at D–D (each representing the number of waves or particles that separately come through S2 or S3) we get a composite curve that has a single sharp peak at the centerline between the two curves; there is no trace of the complex diffraction pattern of C–C.

History does not seem to have a record of the person who first observed this phenomenon, but a loud "Eureka" would have been appropriate. Nevertheless, in the 1920s theoreticians like Bohr, de Broglie, Schrödinger, Dirac and Heisenberg must have been aware of this increase in the wave/particle puzzle since they were all working to explain the "Quantum Paradox."

Modifications of the two-slit experiment deepened the paradox further. Scientists found ways to fire one photon (or one electron) at a time through the slit Sl in the baffle at A–A. If both S2 and S3 were open, photons could go through either slit, but only one at a time. After some thousands of photons had passed through the apparatus, the complex diffraction pattern of accumulated photons at C–C appeared again! As more thousands of photons arrived at the observation screen, the diffraction pattern became more and more distinct. How can single photons arriving one at a time form a diffraction pattern? With what do they interfere? Do they "know" that another photon is coming along in a few seconds? And if they know that, they still cannot communicate or interact with the next photon because they are both traveling at the same speed — the speed of light. The information to form an interference pattern would have to travel faster than the speed of light, which is forbidden according to Einstein's relativity theory.

When photons were sent one at a time toward the slits S2 and S3 with one slit closed, the diffraction pattern disappeared and the smooth bell curve of S2 or S3 appeared on baffle D–D depending on which slit was open.

Although the one-at-a-time photon behavior described above was anticipated much earlier, a scientific group in France actually carried out these experiments in the mid 1980s. A Japanese group performed an equivalent two-slit, one-particle-at-a-time experiment for electrons in 1987–9 (Gribbin, 1995; Baggott, 1992). And finally, in the early 1990s a German group did the same experiments with atoms — in particular helium atoms. All of these experiments gave the same results, if you send photons, electrons, or atoms through a two slit apparatus you get a prominent diffraction pattern on a photographic plate if both slits are open regardless of whether particles go through only one slit or two and a smooth bell curve if only one slit is open. Presumably all other particles behave the same way in an equivalent experimental arrangement.

The Copenhagen Interpretation and EPR Objection

The paradoxes of quantum experimentation led to a variety of interpretations. The first and most widely accepted was Bohr's Copenhagen interpretation or CI. The CI view rests on two main pillars, first, a quantum wave or particle system exists in a state of total indeterminacy before it is measured and second, when measured the indeterminate quantum wave function "collapses" into one of the probability states allowed. This probable state is what we observe when we make an experimental measurement. On a macro scale this is like saying a bush in your garden is not there until you take out a ruler and begin to measure it; then it appears. Because it was originated and supported by an important group of theoreticians, the CI outlook got early support. But many scientists were uneasy because CI closed an interesting new area of study and seemed to imply that human observation makes a quantum system behave (or misbehave) as it does.

In particular, Einstein disliked Bohr's idea that the observer with his measurements somehow creates scientific reality. Einstein was committed to the view that our scientific observations correspond in a direct way with physical reality. The debate between Einstein and Bohr went on for a number of years before coming to a head in 1935 when Einstein, Podolsky and Rosen (the EPR view) proposed an experimental procedure that would distinguish between Bohr's probabilities and Einstein's firm cause and effect reality (Baggott, 1992).

The EPR group suggested what-if experiments that involved two identical particles moving off in opposite directions from an originating source, for instance, an excited atom. Such particles are "correlated pairs" which have opposite wave polarizations (vertical or horizontal) if waves, and magnetic spins (up or down), if electrons. With a correlated pair, if you measure the velocity of one particle accurately, you automatically know the second to the same accuracy. However, Heisenberg's uncertainty principle, which is basic to CI, states that knowing accurately the momentum, or position of one particle precludes knowing the characteristics of the other particle accurately. Because of this apparent contradiction, the EPR group claimed the CI concept was wrong or at least incomplete.

Bohr countered EPR with his usual objection that we know nothing about what is going on within a quantum system until we make a measurement. Since no one knew how to carry out measurements of the EPR type, the discussion rested for the moment. The successful advance of quantum experiments into new areas of chemistry, solid-state physics, and atomic particle theory continued unabated and the CI dependence on the collapse of Schrödinger's probability wave function continued to build triumph after triumph regardless of the conflicting CI/EPR theoretical viewpoints.

During this period, a number of alternates to the Copenhagen interpretation were developed that outflanked CI by employing so-called "hidden variables." David Bohm's is perhaps the most successful. Bohm's model, similar to an earlier de Broglie idea, called on a hidden "pilot" wave to explain quantum behavior. The pilot wave was thought to keep the quantum particle informed about what was going on in its vicinity. Therefore, in the two-slit experiment we can visualize the pilot wave as "informing" the particle or wave whether the second slit was open or not. The quantum particle then "knows" whether it should behave as a particle or as a wave. Scientists are curious people and normally don't like hidden things, but if you can get beyond that, Bohm's theory conformed to the experimental quantum data quite well.

Some serious people offered serious proposals in support of Bohm's idea. Einstein was supportive to some extent because it attributed quantum behavior to real entities not "spooky action at a distance." Unfortunately, Bohm's theory also required faster-than-light signaling between the particle and pilot wave, which is forbidden by Einstein's relativity theories.

Bell's Theorem

The whole area of hidden variables showed how desperate scientists were becoming about the quantum paradox. But, as often happens in science, another door opened onto an entirely new vista when J. S. Bell published his Inequality Theorem in 1964 (Bell, J.S, 1987).

Bell, analyzed the basic mathematical structure of quantum theory as formulated by von Neumann in his authoritative text of 1932. Bell found a loophole in von Neumann's proofs that would allow communication between a quantum particle and an invisible pilot wave. This encouraged Bell to look for firm criteria as to the kind of reality that could support the quantum facts. Using only these facts, without any kind of theory, Bell's inequality specifies the conditions for nature to be "local." Put another way, if nature does not conform to his inequality, reality must be "non-local."

This, in essence, is Bell's theorem. An extremely important aspect of the inequality is that it does not depend on theory in any way. If quantum theory were ever disproved in the future, locality or non-locality would be left unchanged.

Locality and non-locality are concepts dating back to before Galileo and Newton. When you pull a rope to ring the bell in a bell tower, or turn a gear to open a valve, you are performing a local action. A local universe requires all entities to be in some kind of direct contact, including force fields such as gravitation, for an interaction to occur. It has been described as like a clockwork mechanism in which each gear turns another. In a non-local universe, influences (energy or mass) would not necessarily be in observable contact and can spread faster than the speed of light; they are superluminal and apparently spontaneous. A quantum occurrence at position A can instantly influence a correlated quantum that might be light years away at position B.

Bell's inequality or theorem is based on an interesting and surprising bit of pure algebraic logic with the form:

$$(A \times C) + (B \times C) + (A \times D) - (B \times D) = S$$

The surprise is that if you limit the symbols A, B, C and D to the values plus or minus 1, S will always factor out to be plus or minus 2. Taking care that the symbols always have the same value throughout and remembering a little algebraic addition and multiplication, we find that if all symbols are −1:

$$(-1 \times -1) + (-1 \times -1) + (-1 \times -1) - (-1 \times -1) = +2$$

or

$$1 + 1 + 1 - 1 = +2$$

If A only is negative we have:

$$(-1 \times 1) + (1 \times 1) + (-1 \times 1) - (1 \times 1) = -2$$

or

$$-1 + 1 + -1 - 1 = -2$$

All other combinations of plus and minus one also give the same result for S. From this, Bell concluded that if the data from a two-slit type experiment could be assigned to the symbols A, B, C, and D of his inequality, we would know whether CI or the EPR view of quantum behavior was truer (Lindley, 1996).

Bell sided with Einstein in that he thought his inequality would disprove quantum theory if it were ever put to the test. Experimental technique was not up to the task in 1964; it was not until 1972 that John Clauser showed that experimental tests were possible. Finally, in 1982 Alain Aspect and his colleagues at the University of Paris were able to devise an experiment to test definitively Bell's theorem.

Aspect's Experiment

The thought experiments considered by the EPR group involved two identical electrons with opposite quantum "spin" characteristics; however, Clauser and Aspect developed an equivalent procedure that involved polarized light. We can understand the character of polarized light if we imagine looking down the X-axis of figure 1 toward the zero at the left. We would see the electric field component vibrating vertically at right angles to the magnetic component, which is vibrating in the horizontal direction. By convention, a light beam is said to be polarized in the direction parallel to the electric component. In figure 1, for instance, the photon is polarized vertically. The photons in a nonpolarized beam of light have a random mixture of polarization directions but there are a number of ways to polarize light. A simple procedure is to pass the beam through a calcite crystal, which can separate nonpolarized light into two plane-polarized beams, one vertical and the other horizontal. Such oppositely polarized light beams are equivalent to the two opposite electron spins in the EPR formulation. Both spin and polarization have two possible states in a correlated pair of electrons or photons.

Aspect set up an experiment using an excited calcium light source to produce pairs of correlated photons. In this case, correlation means the pair has no overall polarization because the two members of the pair possess opposite polarizations. These photons speed off individually toward two polarization detectors that can be set at various angles. The detectors intercept the photons and determine the polarization direction of each. The correlation of the photons assures the experimenters that if one photon is detected to be vertically polarized, designated $A = +1$, they can be certain the other is horizontal and designated $B = -1$ in Bell's inequality equation.

An additional feature of the Aspect experiment is that the two members of the correlated photons are measured within 20 nano seconds of emission, whereas it would take 40 nano seconds for a signal traveling at the speed of light to traverse the distance between the detector and photon source (in this case about 13 meters). The path of the two photons could also be switched while they were in flight toward the detectors; therefore, detection could be completed before communication between photons and detectors could occur using the velocity of light. Later Nicolus Gisin and his group at the University of Geneva extended the separation between source and detectors to about seven miles with similar results. This showed that the correlation between paired photons did not diminish with distance.

Since there were two correlated photons that could be switched and two detectors, which could be varied in direction, each of the symbols A, B, C, and D could be assigned a unique variable status (vertical polarization A = +1 or horizontal B = −1; switched to right detector C = +1, or to left detector D = −1). A single experiment measured only two variables at a time so that it had to be repeated many times before the results could be averaged to determine values for A, B, C and D in Bell's equation.

Aspect's experiments contradicted Bell's inequality, which requires S to lie between plus 2 and minus 2. The data he collected found S to be twice the square root of 2, or 2.828. Aspect concluded that the measurement of a photon in Paris retained correlation with the polarization of its paired photon any distance away. Such an interaction is non-local. This relationship can be extrapolated to all quantum particles and photons. Since the time of the Big Bang, energy quanta in the universe have been interacting with other quanta in a manner similar to the expulsion of correlated photons by an excited atomic light source as used by Aspect and Gisin (Nadeau & Kafatos, 1999). This can be modeled as a universal quantum entanglement. At the most basic level we know of, the universe appears to be non-local.

If the universe is non-local what will that mean for us? At this time we can only ask questions. Is non-locality a true description of reality? Is non-locality restricted to the quantum realm? Do non-local processes make a significant contribution to living creatures? Will non-locality allow transmission of information at speeds faster than light? Or could we transform a human, for instance, into correlated duplicates and send one off to the galaxy M83 for afternoon tea? Already attempts have been made to extend non-locality to the unnatural vistas of fantasy and science fiction.

In addition, there is that major contradiction between the non-local instantaneous passage of information and the limitation of light velocity to 186,000 miles per second as found by Maxwell and Einstein. However, it has been pointed out that photon correlations occur simultaneously; therefore, no time lapse is involved. The pair correlation is outside of time and distance so it escapes the limitations of speed-of-light velocity.

At least partially, this means that our understanding of the quantum world and probably our everyday world is incomplete. It can also mean our understanding is naïve or that physical reality is much more complex than we can know or visualize. As always, "time will tell" as we search for additional scientific truth.

SCIENTIFIC REALITY

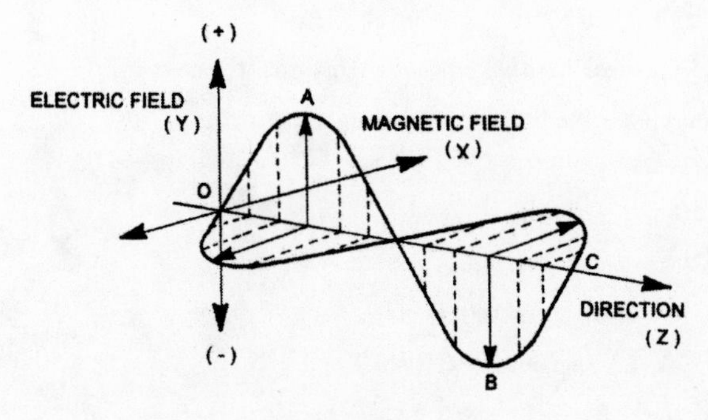

FIGURE 1

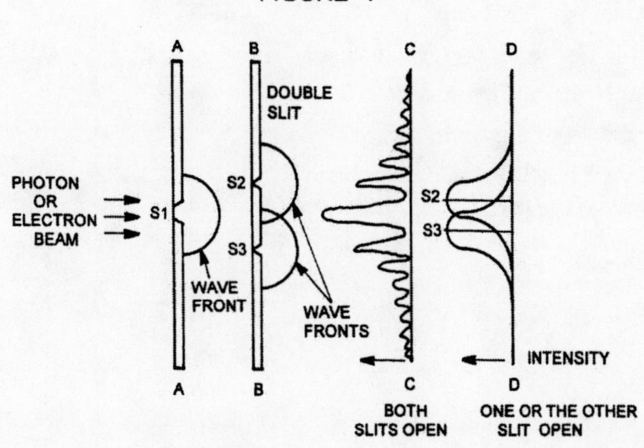

FIGURE 2

Bibliography

Adler, Mortimer 1980, *Ten Philosophic Mistakes*

Baggott, Jim 1992, *The Meaning of Quantum Theory*

Bell, J.S. 1987, *Speakable and Unspeakable in Quantum Mechanics*

Davis, P. and Hersh, R. 1981, *The Mathematical Experience*

Encyclopedia Britannica (1989) (25, 633, 2b)

Greene, B. 2004, *Fabric of the Universe*

Gribbin, J. 1992, *Q is for Quantum*

Gribbin, J. 1995, *Schrödinger's Kitten and the Search for Reality*

Gribbin, J. 1996, *Companion to the Cosmos*

Hersh, R. 1997, *What is Mathematics, Really?*

Kaufmann, W. 1985, *Universe*

Lindley D. 1996, *Where Does the Weirdness Go*

Mayr, E. 2001, *What Evolution Is, Basic Books*

Penrose, R., 2004, *The Road to Reality*

Weinberg, S., 1992, *Dreams of a Final Theory*

Wood, R. W. 1934, *Physical Optics*

Part III. Aspects of Biological Reality

Chapter 9. Darwin and Mendel

Life evolves and the evolution of life is both change and preservation. We are fascinated by the "struggle for life," the "survival of the fittest," "nature red in tooth and claw," natural selection, and extinction. But it is not violent change that is unique to evolution; preservation is the unique factor. We have seen that all things change; the inorganic world of photons, atoms, solar systems, and galaxies are all seething with change — never ending change. If change were the total character of evolution we would still be no more than atoms or molecules. Our lives are momentarily slowed or arrested change built on innumerable, previous changes.

When one species evolves into another, every aspect of the organism does not disappear; rather the preservation recorder, the DNA, incorporates new traits and passes into another species. We can say the former continues to exist as part of the new species; in a sense it is immortal. Evolving life is an accumulation process and a memory of things past.

The concept of evolution did not burst onto the world scene with the publication of Darwin's *Origin of Species* in 1859 as abruptly as it now seems. Information had been accumulating that was not consistent with the then current beliefs of instant creation and fixity of species. The great voyages of discovery of the fifteenth, sixteenth and seventeenth centuries had flooded Europe with new plants and animals, strangers from another world. Some were similar to, or related to, European species; others were not.

Efforts to classify and integrate the new species into the old organizational schemes were not too successful until Linnaeus undertook the problem. In 1735, he published the first of many editions of his *Systema Naturae* which gave each species a binomial name indicating its relationship with other animals or plants. Naming became a popular craze. Specimens poured in from around the world, each submitted with the hope that Linnaeus would use the finder's name or honor another as part of a new species name. This is the source of such wonders as *Nicotiana langsdorffi* (a tobacco plant), *Rana boylei* (a frog), and *Melospiza lincolni* (a sparrow).

During the late seventeenth and early eighteenth centuries, fossil evidence was accumulating showing that earlier plants and animals were sometimes different or absent from the roster of living species. In 1749, the Comte de Buffon published the first volume of his influential *Histoire Naturelle*, a work that was heavily sprinkled with allusions to evolutionary change. The latter were usually covered by careful disclaimers in the hope of avoiding Church disapproval (Eiseley, 1958). Lamarck had proposed a theory in 1809 that acquired traits could be inherited and passed on by the young, leading to the emergence of new species. Among naturalists the belief was growing that species were not fixed but had developed over time from earlier forms.

The Problem with Time

One of the major impediments to the acceptance of evolution was a problem with time. There was not enough of it. Calculations by the Hebrews of the first century BC and the early Christians of the first century AD gave the earth an age of five or six thousand years (Burchfield, 1975). Although most other civilizations had looked on time as extending back into a long indefinite past, European society adhered to the Judeo-Christian view that the universe had been recently created and would soon end, and that all species had been fixed in their present form by a creator.

About 1658, the Anglo-Irish bishop of Armagh, James Ussher, placed the creation of the earth at 4004 BC by adding up the ages of men cited in the Old Testament of the Bible. In 1644, John Lightfoot, Vice Chancellor of Cambridge University, picked 9 AM, September 17, 3928 BC as the beginning of the first day (Dunbar, 1959). Luther calculated 4000 BC as the appointed time. These pundits were in fairly close agreement. However, a total of six thousand years is little time for the evolution of life to its present level of complexity or the observed weathering of the earth's surface. Only violent and catastrophic events could have caused such extensive changes in so few years. Six thousand years is five orders of magnitude less than the Earth's 4.5

and life's 3.5 billion years now posited on the basis of uranium radiometric dating techniques.

Although most geologists accepted the concept of violent change or "catastrophes," Charles Lyell, a Scottish geologist, came to the opposite conclusion; the earth's surface features had emerged slowly by an accumulation of ordinary changes from well known processes such as earthquakes, volcanic eruptions, weathering, erosion and sedimentation (E.B., 1989). His conclusions were partially based on studies of the percentages of fossils from living and extinct species in various levels of rock strata. He found that deeper strata contained a higher percentage of fossils from extinct species. Since deeper rocks must also have been laid down before upper levels, they and their contained fossils were deduced to be relatively older.

Lyle's overall conclusion was that the Earth was probably much older than Bishop Ussher or the catastrophists had ever imagined. He incorporated his ideas and methodology in the three volume work, *Principles of Geology*, published in 1830, 1831, and 1833. Darwin read the work during his famous voyage as a naturalist on the sailing ship *H.M.S. Beagle* from December, 1831 to October, 1836 (West, 1938). Lyell's views on extended time, known as the Uniformitarian model, became a key part of Darwin's evolutionary thinking. Nevertheless, extended time remained a storm center for evolutionary theory until well after firm dating techniques were developed following the discovery of radioactivity by Becquerel in 1896 (Dunbar 1959). Darwin, who died in 1882, never had the assurance of the long period of time his theory required.

Even so, when Darwin published *The Origin of Species* in 1859, the intellectual world of Western civilization was thrown into turmoil. Darwin had marshaled a vast array of disparate information into a coherent evolutionary theory and proposed a mechanism of change that was simple, rational, understandable, and compelling, that is, natural selection.

Natural Selection and Evolution

There is general confusion as to the relationship between the theories of evolution and natural selection. Darwin's original formulation of evolution was a combination of at least five sub theories (Mayr, 1991). These included: 1. evolution in general, 2. common descent, 3. multiplication of species, 4. natural selection, and 5. gradualism. Evolution is the umbrella theory, and natural selection one of its parts. Today there are a number of additional theories supporting evolution as our scientific abilities became ever more sensitive and thorough. These include gene mutation, genetic recombination, reproductive isolation, and the molecular analysis of DNA; all are consistent with both the evolutionary and natural selection theories.

Darwinian evolution is the theory that all living species descended with modifications from earlier species going back to the origins of life on earth. Confirming proof of this is the relatively recent discovery that all life forms from unicellular bacteria to multicellular plants, fungi, and animals use the same molecular DNA composition, structure and general processes of reproduction. The theory of evolution draws on information from all scientific fields for confirmation. Natural selection is difficult to confirm empirically because of the long stretches of time required to observe evolving species. This usually requires time periods much longer than a human lifetime. However, some bacteria and molds reproduce in days or hours and these have been observed to evolve into new species when grown in environments of increasing hostility (Strickberger, 2000; Alberts et al, 1998).

If natural selection should fail for some presently unknown reason, the theory of evolution would still stand since it is based on the total interlocking formulations of all branches of biology including: paleontology, morphology, physiology, embryology, taxonomy, genetics, and molecular biology as well as logic, chemistry and physics. We would simply have to find another process to do what natural selection does now, i.e. provide a mechanism that produces inheritable evolutionary change using the data we now have about evolution. However, as supporting facts continue to accumulate, particularly from molecular biology and fossils, the demise of the process of natural selection becomes vanishingly small. That does not mean the theory will not be modified; it does mean the theory will not be overthrown.

Natural selection is a single but very important part of overall evolutionary theory. It is a mechanism or process that explains how evolution can occur. In brief: individuals vary in their characteristics; no two are exactly alike. Some will be better adapted to their environment at the time and have a better chance of survival and reproduction than others. Darwin explained it in the introduction to *The Origin of Species* in 1859:

> As many more individuals of each species are born than can possibly survive; and as, consequently, there is a frequently recurring struggle for existence, it follows that any being, if it vary however slightly in any manner profitable to itself, under the complex and sometimes varying conditions of life, will have a better chance of surviving, and thus be Naturally Selected. From the strong principle of inheritance, any selected variety will tend to propagate its new and modified form.

Natural selection consists of two main parts: Variation and Selection. Some individuals will be selected because they possess a variation that is beneficial in the environment of their time. But there are also a number of other conditions that are important. There is often pressure on the food supply forcing individuals to compete for food; a surviving individual must reproduce successfully; and the young must inherit the "improved" variation. Also the environment must be relatively stable for a species to continue to exist. If earth temperatures, for instance, had cycled –/+200 degrees during a day, there is little possibility that species like us would be here. If there is a true miracle on earth, it is the relative uniformity and stability of a climate favorable to life over the last 4,000,000,000 years.

Because natural selection as the operative process in evolution is so important, it is worth listing the major observations and facts involved.

1. Individuals of a species reproduce.

2. The young vary from parents and siblings.

3. The environment varies independent of the individuals.

4. Some individuals are better adapted to the environment.

5. Individuals with these variations are more likely to survive. They are in essence selected.

6. The selected individuals must reproduce and pass on the new variations to their offspring by an accurate, reliable inheritance process.

First there are individuals that replicate. The replicated offspring are not perfect copies of the parent or of each other. This is largely due to copying errors, mutations, and crossing over during the meiosis process (cell division). The environment varies independently. Some individuals are better adapted to the particular environment at the moment and survive. These selected individuals must also pass on the better adapted characteristics to their young.

The sequence of events can be summarized as variation within a multiplying species, selection, and reproductive inheritance. All of these steps in natural selection have been verified with massive amounts of evidence. In what follows, we will review some of the high points of this information.

Natural selection is a Darwinian theory that species adapt to their local environment by necessity, i.e. to survive. We speak of species, not individuals, adapting because individuals are limited; they can only adapt within the confines of their genetic makeup. It is a process that challenges every individual, humans included, to live and reproduce in harmony with their surrounding conditions. It is irreversible, impartial, unavoidable, unceasing, aimless, unplanned, and unrepentant (Dennett, 1995). It has been severely criticized from the day of the Origin's publication. Often condensed into the slogan "survival of the fittest," natural selection has been denounced as a tautology, circular thinking, and as not falsifiable; therefore, not scientific (Johnson 1991). All such criticism is false.

Most of the criticism of natural selection comes from those who view it as vicious and abhorrent. How can humans be the product of such a ruthless process? We see the proof of plan and design all around us; how can this be? What are the alternatives? Is it better that some supreme power systematically kills all of its creations after a brief lifetime for an unknown reason? Must we stand in denial of the facts and regularities of nature?

We should note that the man-designed process of laissez faire capitalism works on similar selection principles. Capitalism is a process, which is justified on the basis that the economically successful eliminate all others because they are less efficient. Such an economic process can also be abhorrent, impersonal, exploitive, and vicious. We tolerate capitalism because we have been unable to find another system, which fosters the economic freedom of the individual, is relatively efficient, and also compassionate.

"The survival of the fittest" and "the struggle for existence" which are often used as shorthand for natural selection are not accurate descriptions of the process. Darwin described these phrases as metaphorical only. The phrases are often interpreted to mean that only the fittest survive the struggle. This is far from the case. In a benign environment most individuals survive. Examples abound. It often happens when species are first introduced into an open and benign environment. Consider the rapid expansion of the English sparrow, the European starling, the Japanese beetle and the Arizona house finch when they were first introduced into the northeast U.S in the early 1900s. In the new environment they had no natural enemies and quickly spread. They increased through many generations before they had to compete against an evolving set of predators and enemies.

The most spectacular example of survival within an easy, open environment is the remarkable expansion of the human population over the last few hundred years as mankind occupied new territory and learned to control their environment by improving food production and health procedures.

Although the rate of births is beginning to show a declining trend, presumably due to education and the increasing use of family planning methods, world population will continue to grow for many years because of increasing longevity. Present estimates are that world population will level off at about 9 billion with the fertility rate dropping to a barely sustaining rate of about 2.1 young per female within a few decades. Meanwhile development and exploitation of world resources continues unabated. The balance between these two forces of population/exploitation and earth capacity will determine whether our Earth becomes a Garden of Eden or a steadily declining living habitat.

Bacteria and the Prokaryotes

Scientists have the habit of giving specialized names to anything in nature that seems notably different. A few definitions of terms necessary for an understanding of the role of bacteria in evolution follow:

> Chloroplast: symbiotic, photosynthetic organelles in the cell which originally evolved as individual, separate bacteria.
>
> Cytoplasm: The cellular material within the cell membrane, not including the organelles.
>
> Eukaryotes: (you-carry-oats); Plants, animals and fungi; DNA enclosed in a nucleus; many organelles; generally multicellular, bisexual and with larger cells.
>
> Meiosis: Special sexually reproducing cells divide to form sperm and egg cells, each with only one of the paired chromosomes normal to each somatic or body cell.
>
> Metabolism: the energy process leading to the growth and maintenance of living cells.
>
> Mitochondria: organelles that produce chemical energy in the form of ATP (adenosine triphosphate) by oxidizing carbohydrates; consume oxygen and release carbon dioxide; evolved symbiotically from once independent bacteria.
>
> Mitosis: The cell division process in which chromosomes split lengthwise then replicate. After cell division each of the twin daughter cells has a copy of the original paired chromosomes. This is the body (somatic) cell division process.
>
> Organelles: specialized functional bodies within the cell such as chloroplasts, nuclei, and mitochondria.
>
> Protista (or Protists); single-celled organisms — yeast to amoebas.
>
> Prokaryotes: (pro-carry-oats); Bacteria; without a nucleus, DNA floats freely in the cellular cytoplasm; mainly single celled but some form chains or clusters; few organelles.
>
> Symbiosis: A close interaction of two individuals from separate species, which usually benefits both. Examples — lichens, chloroplasts and mitochondria.

The living universe can be divided into creatures with cells that have nuclei and those without nuclei. The prokaryotes have no separate wall around their DNA, few organelles, and consist mainly of single celled bacteria. The cells of eukaryote species have enclosed nuclei that contain the chromosomes. They are mainly multicellular and bisexual species that have numerous organelles and are represented by plants, animals and fungi. When we speak of nature we are usually referring to animals and plants while bacteria have been largely hidden from our vision.

We can begin the story of life's evolutionary history with an extremely hot Earth about 4.5 billion years ago. The Earth had coalesced from debris left after the sun formed by gravitational forces. The Earth rapidly condensed into land, sea, and underlying rock layers with a nickel-iron core. The atmosphere was chemically reducing, that is, anaerobic or lacking any significant amount of free oxygen and composed largely of hydrogen, nitrogen, ammonia, hydrogen sulfide, water vapor and carbon dioxide gases. When combined with energy from solar radiation, lightning discharges, radioactive decay, or high volcanic temperatures, this inorganic gaseous mixture can produce a great variety of small organic compounds such as methane and amino acids.

Over time, simple, carbon based molecules formed into more complex molecular structures that could replicate themselves from surrounding materials. Replication sometimes yields variations since replication is seldom a perfect duplicating process. Natural selection could now begin — replicas with variations that are well adapted to their surroundings are more likely to survive than those replicas that are not well adapted.

The replicating molecules developed enclosing membranes which allowed them to concentrate selected materials and formed the first living single cells. Cells have been the basic structural unit of life ever since; they are the first prokaryotes — the protocells that became bacteria. The earliest energy sources for metabolism came from small organic molecules or from sunlight and inorganic compounds such as hydrogen sulfide — probably the first form of photosynthesis. The purple and green sulfur bacteria of today actually use this process to form glucose (Atkins, 1991).

$$CO_2 + 2H_2S = (CH_2O) + H_2O + 2S$$

CH_2O symbolizes the ratio of carbon, hydrogen and oxygen atoms in a carbohydrate; glucose, for example, is $C_6H_{12}O_6$.

Bacteria (the kingdom *Monera*) are single-celled creatures that reproduce by dividing into two identical daughter cells. In size they are about 1–10 microns (a micron = 10^{-4} cm) and require a powerful optical or electron microscope to observe in detail.

The single-celled bacteria and their replicating predecessors, were alone for at least the first 1.5 of the approximately 3.8 billion years of life on Earth. They are the most enduring form of life known and make up more than fifty percent of the biomass of Earth even today. During this dominant period, bacteria developed the basic cellular processes that later made more complex life forms possible. Yet the paramount role of bacteria was largely unnoticed until relatively recently.

The achievements of bacteria began when they enclosed their replicating molecules and cytoplasm within a membrane. Over time, specialized kinds of bacteria, cyanobacteria for instance, evolved that could combine the energy of the Sun's photons with water and carbon dioxide to form carbohydrates and oxygen. This photosynthetic process, which required the molecule chlorophyll, is the source of all organic compounds available on earth today including the energy sources: oil, natural gas, coal, wood, and, most important, our food. The chemical notation for all these benefits is:

$$CO_2\text{ (gas)} + 2H_2O = (CH_2O) + H_2O + O_2\text{ (gas)}$$

From this equation it appears that carbon dioxide reacted with two molecules of water to form carbohydrate and released its oxygen atoms as a gas molecule. However, experiments using water containing heavy isotopes of oxygen proved that the oxygen came from water. This so-called "water splitting reaction" is the source of the oxygen we breathe and use in metabolism as well as the hydrogen in carbohydrates. It is estimated that 45 trillion pounds of carbon in the form of carbon dioxide are processed each year by photosynthetic bacteria and the chloroplasts of green plants to yield oxygen and carbohydrates (Atkins, 1991).

During the era of bacterial dominance, photosynthetic cyanobacteria slowly began to increase the atmospheric oxygen level from less than one percent to near the present level of 21 percent. The importance of this "oxygenation" of the atmosphere can hardly be overstated. It occupies the period from roughly 2.5 to 1.6 billion years ago or from the era of complete dominance by the prokaryotes to cohabitation with the eukaryotes, their present companions. It also indicates the enormous changes that living creatures can have and still have on the physical conditions of earth.

Oxygen was a hazardous intruder into the universe of life. The chemical reactivity of oxygen is well known from its ability to combine with, oxidize, or burn almost anything. The earth is largely composed of oxides of silicon, aluminum, iron, calcium, sodium, etc. When such elements react with oxygen, the resulting compounds have less free energy and therefore are more chemically stable than the original elements. In effect the elements are removed from further chemical activity. These oxidation reactions were simply conforming to the requirements expressed in the Second Law of Thermodynamics.

As the level of atmospheric oxygen increased, living creatures experienced an environmental crisis of the first order. Originally living creatures evolved under reducing conditions and now were forced to adapt to an atmosphere poisoned with free oxygen. Even the cyanobacteria that were producing the oxygen were subject to damage. However, the oxygen increase occurred slowly over hundreds of millions of years so that each living generation needed to make only a slight adjustment. Nevertheless, all species were faced with evolutionary pressure to change in the direction of increased tolerance to oxygen.

There was one overriding advantage in the new situation; oxygen metabolism is much more efficient than the older methods, which includes fermentation and hydrogen sulfide photosynthesis. Compared to fermentation, oxygen metabolism can be as much as fifteen times more efficient (Margulis, 1986–1997). This leap in metabolic energy efficiency increased bacteria's potential for the evolution of new functions such as mobility, the ability to move about in the sea with cilia or tails, and sensory functions that could give movement direction.

Bacteria's growing adaptation to oxygen metabolism led to the emergence of the eukaryotes about 1.8 billion years ago. Perhaps the eukaryote nuclear membrane evolved to protect the precious DNA from oxidation at that time.

Earlier, we described bacteria as reproducing by dividing into two identical daughter cells; unfortunately nature seems to delight in exceptions, particularly if they confound us. When bacteria die and decompose, they can leave segments of DNA floating in the surrounding media. These fragments can be directly transferred through the cell wall of other living bacteria (Alberts et al, 1998). This is a process called transduction, which complements the process of conjugation or the essentially sexual transfer of genetic material by direct contact between bacteria. The latter process can occur only when one of the bacteria has a specific organelle — a plasmid able to form a cytoplasmic tube through which the genetic material can flow.

Possession of three methods of reproduction — division, transduction, and conjugation, gave bacteria an increased genetic adaptability. The DNA fragments were potentially available to fill gaps in bacterial defenses against hazardous changes in the environment. There was less need for the common incremental evolutionary processes to develop over many generations. Laboratory experiments have clearly shown that genetically variable bacteria can adapt to abrupt changes in conditions that quickly kill bacteria without such genetic reserves. Easy genetic interchange gave bacteria common information and abilities. They had many of the characteristics of a single species related in a community of common interest (Alberts et al, 1998; Margulis and Sagan, 1986–97).

The Eukaryotes and Symbiosis

The earliest eukaryote fossils date from about 1.8 billion years ago compared to the earliest fossil date of 3.5BYA for prokaryotes that were bacteria (Schopf, 1999). There is little doubt that eukaryotes evolved from the much earlier prokaryotes; the question is how?

Eukaryote cells are generally much larger and complex than prokaryotes; their DNA is enclosed within a nuclear membrane compared to the prokaryotes' free floating DNA. Prokaryotes reproduce by splitting in two, often growing a bud which enlarges and splits off as an identical daughter cell. Eukaryotes reproduce by a much more complex dual process; body cells reproduce by mitosis, in which the chromosomes are split lengthwise to form independent, identical cells; the sexual cells are formed by the meiosis process from special relatively primitive cells set aside early in an individual's life. These split into sperm or egg cells, each with one half the number of DNA strands in an adult cell and then recombine during the sexual process to form a fertilized cell that can continuously divide and grow by mitosis into a normal adult. In the human adult, the number of chromosomes is 46, the egg and sperm cells each contain 23 and when they unite, form the first cell of an adult with the required number of 46 chromosomes.

In the past fifty years, scientific evidence has accumulated showing that many of the organelles now in eukaryote cells come from the symbiotic interaction of bacteria with a host eukaryotic cell. The evidence is particularly convincing for two organelles, Chloroplasts, which by photosynthesis convert water, carbon dioxide and photons of light into oxygen and carbohydrates and Mitochondria that produce chemical energy by oxidizing carbohydrates.

The evidence for symbiosis is based on a variety of factors. First, mitochondria and chloroplasts are similar in size and shape to bacteria. Second, they have DNA and reproduce themselves independently of the host eukaryote's DNA reproductive cycle. Third, they reproduce by the bacterial method of budding and division into two identical daughter cells; they lack the meiosis process. Fourth, mitochondria's genetic makeup resembles DNA from contemporary bacteria that produce energy by the oxidation of carbohydrates. And fifth, the DNA of chloroplasts resembles that of today's cyanobacteria rather than the DNA of the eukaryote host cell (Margulis, 1998; Ryan, F., 2002).

There remains little doubt that most plant and animal cells contain remnants of once independent bacteria that were consumed by, or penetrated the original host cells. The chloroplasts that are present in green plant cells are now believed to have originated as independent photosynthesizing bacteria that became enclosed within other cells symbiotically; similarly, mitochondria were once independent bacteria that have accommodated themselves to a comfortable life within their host cell.

Symbiosis can be broadly defined as the relationship of two species that live together with a result that is mutually beneficial but not necessarily so (E.B., 1989). Some relationships are parasitic, such as the hookworm that can live in the intestinal tract but causes anemia in the host. Others are marginal, like the shrimp that "clean" the moray eel's teeth. The shrimp, of course, are looking for scrapes of food and the eel seemingly doesn't mind. But many symbiotic relationships are of extreme importance to the evolutionary path that has led to the present.

The concept of symbiosis had its beginnings in the 1860s when botanists became aware of the strange characteristics of lichens. They had been variously classed as algae, mosses or fungi when in 1868 the Swiss botanist, Simon Schwendener, described all lichens as associations of a fungus and an alga. He described them as in a master–slave relationship with the fungus surrounding and holding the alga as an exploited vassal. The idea was repugnant to both systematists and botanists. However, as more investigators studied lichens it became clear they were in fact combinations of fungi and algae. The relationship was beneficial; the algae contributed photosynthesis and carbohydrates, the fungi water and minerals (Sapp, 1994).

Symbiotic relationships are wide spread and take many forms. For instance, some species of ants in South America and termites in Africa cultivate fungus in especially prepared compost plots. The fungi are more efficient at breaking down fibrous materials and provide the insects with better food than their own stomachs. Some ants also care for "herds" of aphids, which are able to suck the juices out of plants. The ants stroke the aphids to "milk" them of sweet excreta (Dawkins, 1976). Many legumes have a symbiotic relationship with nitrogen-fixing bacteria growing on the roots providing the legume with a soluble nitrogen fertilizer (Gould, 1996).

There are those who believe that symbiosis is a substitute for natural selection or, at least, facilitates evolutionary change. The latter argument is probably true since "facilitates" allows a wide interpretation but the former argument is without a factual base. Mitochondria and chloroplast bacteria undoubtedly took many hundred million years to evolve their energy producing and photosynthetic capabilities. That eukaryotes use these capabilities shows that nature can take advantage of opportunities that come along by chance. Whether eukaryotes would ever have developed their own form of photosynthesis and energy conversion is problematic.

The Evidence for Evolution and Natural Selection

All the fields of science enumerated a few pages back lead to and confirm the theory of evolution. We shall briefly review some of this evidence and then dwell in more detail on the fossil record and genetics.

Embryology, Morphology, and Physiology

There is a remarkable similarity in embryonic development among species that corresponds to their evolutionary stage. The process is often described as follows: the embryonic development of the individual recapitulates the emergence of the vertebrate classes. Embryologists have long observed that fish, amphibians, reptiles, birds and mammals have gill-like structures in the embryo stage, yet only fish and amphibians actually use gills in the adult stage. The heart shows similar transitions. The fish heart has two chambers, the amphibians and reptiles have three, and birds and mammals have four. During embryonic development bird and mammal hearts pass through all of these stages in the order given i.e. in the order of the evolutionary appearance of the class of species.

There are also vestigial structures in adults that have no essential function. Humans often have their appendix removed with beneficial results. Yet for rabbits the appendix is a necessary part of the digestive system. Arms, wings, legs, and flippers have many corresponding bones that show an evolutionary relationship. Forelimbs evolved into wings in birds and bats.

Homologous functions are widespread. It can hardly be chance that all green plant species make use of chlorophyll to convert the sun's energy into a chemical form of energy needed for internal use. DNA, of course, is the most dramatic example of a single structure and process in use for the same function in all species.

Similarities of function, structure, and development are widespread in all forms of life. The only theory that rationally explains these similarities is the Darwinian theory of evolution.

Domestication as Selection

Darwin was born and raised in Shrewsbury, Shropshire — a rural part of England bordering Wales. As a youth, he was well aware of the plant and animal breeding activities of the local farmers. At that time it was general practice to mate only the most vigorous and productive individuals for breeding the next generation of plants and animals. This was only common sense — fast horses usually produce fast colts. This is so obvious that the earliest humans must have observed it and selectively bred animals in prehistoric times. In the England of the 1800s, selective breeding of livestock was a way of life. Exhibitions, shows and competitions allowed breeders to show off the advantages of their particular strain of livestock or fowl. A common subject of English art during the period was prize horses, cattle, sheep, chickens, gamecocks, geese and ducks. The advantages of selective breeding were known to all. A natural deduction from this knowledge is that species are not fixed; they can be made to change, to evolve. Darwin's mind was exposed to these ideas of change from an early age.

After his voyage on the *Beagle* and convinced that living forms did indeed evolve, Darwin's problem was to find a process in nature that could bring about a natural increase of beneficial traits. On September 28, 1838, as he was reading Thomas Malthus's 1798 "Essay on the Principle of Population," the twenty-nine year old Darwin had a flash of insight into the mechanism of evolution (Mayr, 1991). Malthus argued that man's desire for endless improvement in the human condition was hopeless because population increases will always outgrow food production. Population increases geometrically or by multiples of 1, 2, 4, 8, 16..., whereas food production increases arithmetically or by multiples of 1, 2, 3, 4, 5. Darwin realized that all creatures in nature faced the same problem and "a struggle for existence" would occur. He also extended the concept to all the vagaries of the internal and external individual environment. Here was a natural selection process that would do what human selection did for the modification of domesticated plants and animals.

Over the years Malthus' argument has been severely criticized. After all, the human population has increased enormously since 1800 without much famine. Such criticism ignores the fact that there have been many famines in such areas as China, India, Africa, Russia and Ireland over the last two hundred years. Westerners have also expanded into newly discovered continents, relieving population pressure for the time being. This is a unique event that will not occur again. In nature it is common for species to encounter famine; the Grants (1989), who studied the finches on a Galapagos island, determined that finches experienced famine periodically depending on the yearly rainfall and the consequent seed supply. The advance of science has greatly increased crop yields. But now there is no more land to exploit; the agriculture successes of science are slowing; and population still increases. We might soon be forced to dam up the seas and level the mountains.

Bacterial Resistance to Drugs

Diseases like tuberculosis and malaria are becoming increasingly resistant to drug treatment. Gonorrhea has become so resistant to penicillin that the disease is becoming a potential contagion. The reason for this increasing resistance is that bacteria must be treated with overwhelming amounts of antibiotics to be effective. If they are treated with less, the few bacteria that survive have in effect been selected for their drug resistance. These propagate to become the dominant disease agent. Repeated use of a drug in this way quickly leads to new species of bacteria that are immune to penicillin, for instance. The resistant bacteria are new species because their genetic makeup has been significantly altered.

Similar results can be obtained in a laboratory test tube by treating bacteria with increasingly harsh chemicals; then repeat the treatment on the survivors again and again. Such "test tube evolution" can also select new molecules that are able to perform new biological functions.

Agricultural chemicals are also losing their potency for similar reasons. Pesticides, fungicides, and herbicides are becoming less affective forcing farmers to use more complex and costly chemicals to combat increasingly resistant insects, fungi, and weeds.

Paleontology and the Fossil Record

The first hard, physical evidence for the evolution of life came from the fossil record. In general, fossils are traces of the activities or body remains of creatures from the remote past. Even though the world is littered with fossils of all kinds, the ancients seemed to have found few of them. About 450 BC Herodotus and Xenophanes described imprints of fish and seaweed in the quarries of Syracuse and Malta, but thought little of them. Empedocles ascribed huge bones found in Sicily to a race of mythical giants (Tassy, 1991). Otherwise, the ancient records are largely silent. Later, as mining became common, more fossils were uncovered and by the 15th, 16th, and 17th centuries the opinion was expressed by many, including da Vinci, Leibniz, and Buffon, that fossils were the remains of ancient creatures (E.B., 1989). However, it was George Cuvier who gave the term fossil its modern meaning in about 1800.

Cuvier fathered two new branches of science, comparative anatomy and paleontology. He insisted that every part of an animal's body must be consistent and correlated: if it eats grass, it must have teeth suited for grinding; if it is a carnivore, it must have fangs to grasp and tear flesh. He showed that comparisons between skeletons, part by part, will lead to a coordinated and rational whole.

Using this comparative model, Cuvier was able to reconstruct an animal from the evidence of a few fossil bones, a method used by paleontologists ever since. He assembled several complete fossil skeletons from scattered fossils and showed they were not species known to man and were actually extinct. The study of fossils led on to the arrangement of extinct species into related groupings and detailed comparisons to modern forms.

Fossils

Although fossils are widely scattered, the process of fossilization is a complex and relatively rare event. Dead plants and animals are devoured and scattered quickly by scavengers or decomposed by bacteria. The total mass of all the organisms that have lived since the Cambrian period (about 500 million years ago) has been estimated as equal to the total mass of the earth (E.B., 1989). This means the atomic materials, carbon, hydrogen, oxygen, phosphorus, etc., of plants and animals have been recycled many times. Since only about 200,000 fossil species have been found and described, there are undoubtedly millions more to be discovered.

The conditions for fossilization are stringent. First, the dead individual must be protected quickly from oxidation, decomposition and scavengers. This can occur by sedimentation along the margins of lakes and seas, or encapsulation by tree sap to form amber, etc. Over time, additional layers of sediment can compress the remains into shale. If there are acidic waters present, the material of the organism can be replaced by silica, calcium carbonate, or iron oxides. This occurs so slowly and delicately that often every detail of the specimen is duplicated. Casts or molds can form if the organic remains dissolve out of the compressed sedimentary rock. Shells, bones, and other hard parts can be retained largely unchanged and Arctic cold or tar pits preserve ancient bison, mammoth and rhinoceros almost intact.

To date, the earliest fossils are from the Apex cherts of northwest Australia. They were found layered between two beds of volcanic lava, which were dated quite precisely at 3,458 +/−1.9 for the upper and 3,471 +/− 5 million years ago for the lower layer. Therefore the Apex cherts must be about 3,464 million years or 3.4 billion years old. This is only about 500 million years after the estimate that life began about 3.8 billion years ago. The eleven fossil bacteria found were studied microscopically using thin sections of the chert minerals. Three of the bacteria species were remarkably similar to modern cyanobacteria. This implies that the three were photosynthetic and that the atmosphere already contained significant amounts of oxygen (Schopf, 1999).

There is other evidence of early life from the Cambrian and Precambrian epochs. Black cherts from the Gunflint formation of the northern slope of Lake Superior contain black filamentous traces of organic matter. These have been dated at 2.1 billion years. And fossil multicellular jellyfish have been found in rocks from the Ediacara Hills of South Australia, dating from 0.6 billion years ago.

The evidence of the fossil record has been criticized, as full of gaps with many missing links. This is an odd argument since there must always be gaps between species. By definition, members of a species have characteristics that are different from other, similar species; therefore there are always gaps between them. As the paleontologists continue to work, the size of the "gaps" are steadily decreasing.

The reasons for the relative scarcity of fossils can be summarized as follows:

1. Good conditions for fossilization are rare.

2. By definition, a new species has few members. Initial populations are small, grow slowly and occupy a small area.

3. Many differences between species are internal and made of soft tissue. Often these cannot be fossilized. Before the Cambrian Explosion of about 500 BYA living creatures had few if any hard parts such as bones or shells.

4. Even well formed fossils wear out. Chemical action, geological changes, erosion, etc. all take their toll.

5. Only a fraction of all fossils have been found.

Darwinism has answered the major questions concerning fossils. Why are there any at all? Why does the fossil record show increasingly simpler forms as we go back in time etc.? All such questions are convincingly and rationally answered by the theory of evolution.

Genetics — Mendel's Peas

Genetics, the science of the inheriting mechanisms of living creatures, took its first major step forward with the experiments of Gregor Mendel (1822–1884), beginning about three years before Darwin published *The Origin of Species*.

Mendel's early life was similar in several respects to Darwin's. He was raised on his parent's orchard-farm in a rural area of what was then the Austro-Hungarian Empire, where the apple harvest or the weight of hogs were more important topics of conversation than the affairs of empire. He had an abiding interest in the breeding of plants and animals and in natural science. In 1843 Mendel entered the Augustinian monastery at Brunn (now Brno, about 100 miles SE of Prague) and in 1847 was ordained a priest. He studied mathematics and sciences at the University of Vienna from 1851 to 1853 and after a period of teaching was elected abbot of his monastery in 1868. Like Darwin, Mendel shared the boyhood advantages of a rural upbringing, interests in farming, plant breeding, and natural history, an average academic record, and formal religious training. Mendel, however, experienced nothing like Darwin's unique five-year voyage as a naturalist on the sailing vessel H.M.S. Beagle.

Beginning about 1856 Mendel began the experimental studies that led to his discoveries on the inheritance of physical traits in pea plants. His experimental equipment consisted of a strip of earth in the small monastery garden at Brunn, a few bags of peas, curiosity, discipline, a record book, and persistence.

To understand the problems both Mendel and Darwin faced, we should consider the general state of knowledge about the inheritance of plant and animal traits in the middle years of the 19th century. Darwin had long speculated on the mechanism of inheritance, an undeveloped part of his evolutionary theory. He at first favored the idea of "blending" inheritance: a plant with red flowers when crossed with another variety with white flowers would have offspring with pink flowers. An engineer named Fleeming Jenkin convincingly proved that diluting a population possessing red traits with a single white trait would have minimal effect and quickly disappear much like a drop of black ink dropped into a clear lake. Darwin then proposed the existence of invisible particles that were produced by organs and tissues and concentrated in the sperm and eggs before fertilization. These he called "gemmules" (pangenesis). The idea of separate particles was correct, but we know now that every cell has a double set of particles, i.e. the genes on the chromosomes which control the inheritance and development of every new individual.

Although Hooke had discovered the cellular nature of all living tissue by microscopically examining a thin section of cork as early as 1665, most observers thought the cells contained only a clear liquid. Improved microscopes and chemical dying techniques developed in the 1870s allowed chromosomes, nuclei, and cell details to be visually observed (Edey and Johnson, 1989). Eduard Strasburger described the cell process of mitosis (cell division) from microscopic studies of plant tissue (Nordenskiold, 1935). But none of this was available to Mendel or Darwin in the 1850s.

For the sake of simplicity and clarity in a discussion of the inheritance process, we must leapfrog Mendel for the moment to consider what we now know about the basic cell processes of mitosis and meiosis (Villee & Dethier, 1971; E.B., 1989). All the body cells of plants and animals contain a nucleus surrounded by a membrane. The nucleus contains very fine threads that at the appropriate time become rod-like chromosomes. Every species has a definite number of chromosomes. In humans that number is 46; Mendel's peas contain 14. But the chromosomes are actually in pairs. One of the pair comes from the maternal parent, the other from the paternal. This means there are really only 23 different chromosomes controlling the inheritance of humans. To avoid confusion of nomenclature we will refer to the chromosomes as single, or paired. Each chromosome of a pair contains a gene or DNA complex that can control a trait such as tallness or shortness. Here is where the phenomena of dominance vs. recessive appears. Although there are two genes for each trait, only one dominant gene is needed to exhibit the trait. In the case of Mendel's pea plant, tall is dominant, a single tall gene suffices to create a tall plant; a short plant will appear only if both genes of the chromosome pair are for short plants.

Mitosis is the cell division process that carries out the development and maintenance of body tissue and organs. It is by far the most common cell division process.

Although the process of mitosis is continuous, it goes through several phases. As the process begins in humans, the thin threads of DNA duplicate themselves within the nuclei making the number of chromosomes effectively 92. These 92 are called chromatids. Next the threads of DNA contract and thicken into short bent rods; the nuclear membrane disappears and the chromatids divide lengthwise, each half moving to an opposite side of the cell. A membrane forms around each half creating two new nuclei with 46 paired chromosomes each. Meanwhile the chromosomes return to the elongated tangled thread form of functioning DNA. A furrow develops between the two nuclei forming two new cells. Mitotic cell division into two identical daughter cells is now complete. In humans the total process takes about 24 hours, although some cells require a year or more. In the early divisions of a fertilized egg cell the process can take 30 minutes or less.

The second cell division process, meiosis, involves only a small group of primitive, undifferentiated cells in adults that are set aside very early in an individual's life. During meiosis these cells undergo changes that produce the sperm cells of the male or the egg cells of the female. Meiosis follows the mitosis process with two very important additions. First, meiosis is complete after two chromosome divisions rather than one as in mitosis. The result is that sperm and egg cells contain only a single set of chromosomes or 23 in the case of humans. The first division reduces the count from 92 to 46 and the second from 46 to 23 single chromosomes. At fertilization the set of 23 from the egg cell combines with the set of 23 from the male sperm cell to form the first paired 46 chromosome cell of the new embryo. From this point on the cells will divide by mitosis until the adult stage when again a few previously set aside cells will undergo meiosis to form the reproductive egg or sperm cells with 23 chromosomes each for the next generation.

The second major addition to meitosis to yield the meiosis division is the crossover process. Early in meiosis there is a period called synapsis in which the entwined chromosomes can interchange genetic material (Villee and Dethier, 1971). This is a complex process of rearrangement of the genetic material on the chromosomes involved. The importance of this step can hardly be exaggerated since it provides a large measure of the variability characteristic of living creatures and helps make each individual unique.

Mendel knew there were 'pure' strains of pea plants that always produced a definite trait, for instance tallness, no matter how many generations were inbred. He decided to determine if pure strains of plants, when crossed with other pure strains produced mixed or pure offspring. He studied thirty two traits of the pea plant for two years before selecting seven pairs of pure strains to cross, including tall versus short plants, white versus red flowers, smooth versus wrinkled seeds, and yellow versus green seeds. These first crossed generations were then crossed with each of their siblings to produce the second generation. The results were startling.

One of Mendel's experiments bred a pure tall plant with a pure short plant. The seeds of this cross produced only tall plants! What happened to the short trait? Giving a name to an anomaly is perhaps the first step in understanding it. Mendel called the tall trait dominant over the short trait which consequently was called recessive. The recessive trait was not destroyed or missing; it was hidden. Using more recent terminology, the hidden trait was still on one of the paired chromosomes, i.e. in the genotype. The genotype encompasses the total genetic structure of an individual, including both dominant and recessive genes on all chromosomes. The phenotype, on the other hand, is the total of characteristics exhibited by physical appearance and life functions as determined only by our dominant genes. When we look in the mirror we see our phenotype, which is the physical expression of the dominant half of our total genotype.

After Mendel crossed the first generations with each other, his problems compounded. In this second generation he found there were both tall and short plants again. His years of studying mathematics, however, alerted him to a difference; there were about three tall plants for every short plant. He repeated the experiment and then studied the other paired traits. The results were always the same; there were about three showing the dominant trait for each plant showing the recessive trait. Mendel's results were based on observations of each trait for literally thousands of pea plants, making this the first use of statistics in the analysis of a biological problem. It was a painstaking process as well; each cross-fertilization required Mendel to pry open the first pea plant's flower bud, remove the anthers to avoid self-pollination, then dust the stigma with pollen collected from the second pea plant.

To understand how Mendel's 3/1 ratio came about, it is necessary to use a symbolic notation in order to keep track of the individual traits. We use brackets to enclose upper and lower case letters symbolizing the particular traits of interest within a chromosome pair. The upper case letters stand for female traits and lower case for male traits. T and t stand for tall plants and S and s for short plants. For instance, (Tt) stands for two dominant tall traits from the male t and female T lines of pure tall plants and similarly (Ss) for two recessive short traits from short plants. To demonstrate what happens in the first cross of pure lines we first must go through the process of meiosis to form the sperm and egg cells and then consider all the possible crosses of these reproductive cells when fertilization occurs. Symbolically we have:

1. (Tt) → meiosis → (T) + (t) the reproductive cells
2. (Ss) → meiosis → (S) + (s) " " "

If we cross each (T) or (t) cell of line 1, with each (S) or (s) cell of line 2, the possible combinations are four:

3. (TS) + (Ts) + (tS) + (ts)

All the seeds of these crosses will produce tall plants since they all have a dominant T or t in at least one of their chromosome pairs. This is what Mendel had found and used in his dominant/recessive classification scheme. Now for the second generation; if we put the four pairs of line 3 through the meiosis process, we have:

4. (TS) → meiosis → (T) + (S) the reproductive cells
5. (Ts) → " → (T) + (s) "
6. (tS) → " → (t) + (S) "
7. (ts) → " → (t) + (s) "

The combinations of the egg and sperm cells of lines 4 to 7 can be determined by taking any four reproductive cells and crossing them with the remaining four in all possible combinations. For instance, crossing the reproductive cells of lines 4 and 5 with those of lines 6 and 7 we find sixteen possibilities:

8. (Tt) + (TS) + (Tt) + (Ts) + (St) + (SS) + (St) + (Ss) + (Tt) + (TS) + (Tt) + (Ts) + (st) + (sS) + (st) + (ss).

For this second generation there are 16 possible crosses; 12 have a dominant t or T gene and are therefore tall; the remaining 4 have two recessive s or S genes only and are therefore short. The ratio of tall to short is 12 to 4, or exactly the 3 to 1 ratio Mendel had found earlier.

Mendel reported the first results of his researches to the forty members of the Brunn Society for the Study of Natural Science on a cold February evening in 1865. A month later, he elaborated his new formulation of the inheritance process. According to the minutes of the meetings there were no questions, no comments. The Society published the lectures in 1866 and sent them to various universities and learned societies throughout Europe. There was total silence. Not quite; Mendel contacted one qualified person, Karl von Nageli, a renowned botanist; he sent him a printed copy of his paper with a letter asking for comments and perhaps collaboration on a study of the hawkweed, a known favorite of Nageli's.

Unfortunately hawkweed, we know now, is one of those common examples in biology where rules are made to be broken. Hawkweed has very small flowers and is almost impossible to fertilize with certainty; it also reproduces sometimes without fertilization and is generally very variable. Mendel worked on hawkweed for several years, but the collaboration with Nageli never bore fruit. His administrative duties concerning disputes over taxes on the abbey gradually dominated his activity. He died in 1884 without the slightest recognition from the scientific community for his achievements. There is nothing to suggest that Darwin ever heard of Mendel, but Mendel must have read *The Origin of Species* because of its fame (L. Eiseley, 1958). Apparently Mendel never made an attempt to contact Darwin about a subject of so much interest to both of them.

The overall result of Mendel's work was that it documented the first proof that the hereditary traits of plants come in discrete paired elements (now known as genes in the chromosomes); there is no blending. Traits are inherited independently of each other and can be dominant or recessive. The 3 to 1 ratio was the first instance of a mathematical method applied to such internal processes as inheritance in living creatures. From this point on, biological investigations would become quantitative as well as qualitative and descriptive.

But where were these discrete paired elements? In 1865, no one knew that the nucleus contained chromosomes, the chromosomes contained genes, and that genes were made of atoms and molecules. It was not until the 1870s that chromosomes were first observed, and much later that genes and molecules were detectable.

Mendel was somewhat fortunate in his choice of pea plants. The traits he chose were on separate chromosomes that made the evidence of cross fertilizations unambiguous. Under some circumstances, traits can seem to blend, i.e., red and white will form pink flowers. But this is a case where red and white traits only partially assert themselves and the offspring can revert to red and white flowers again. This is yet another example that, in biology, rules are made to be broken; there are few absolutes.

Unfortunately, Mendel's work lay unrecognized from 1865 until 1900. Darwin was never able to use it to solve one of his major problems: how were changes that passed the test of natural selection inherited? In 1900, three scientists, Hugo de Vries, Carl Correns, and Erich Tschermak, working separately on plant variation and hybridization, discovered Mendel's work and reported they had duplicated his results. Very quickly the great value of his findings became known and led to a burst of activity in the new field of the gene, genetics.

One would think the rediscovery of Mendel's laws would lead to a reinforcement of Darwin's theory, but that was not the case. In studying evening primroses, De Vries found sudden, dramatic, and inheritable changes in a few plants. He called these mutants or mutations and convinced himself and many others that mutations were the major vehicle through which evolutionary change occurred. The slow Darwinian process of variation and selection seemed superfluous. Darwinism reached its lowest ebb during this mutationist interlude. But again, it was soon shown that the traits thought to be mutations were actually hidden variables within the evening primrose's normal genotype (Edey and Johnson, 1989).

Meanwhile, at Columbia University, T. H. Morgan began his famous studies of the fruit fly, *Drosophila melanogaster*. Morgan was determined to find the source of individual variation since he was convinced Lamarck's, Darwin's, and Mendel's ideas were wrong. He fixed on the new notion of mutations as the most likely place to begin and exposed his flies to severe conditions including x-ray irradiation, cold, heat, and strong chemicals in the hope of inducing mutations. He gave up after two years when no mutants appeared. Unexpectedly, in 1910, a natural mutant, a white-eyed fly, spontaneously appeared among the normally red-eyed Drosophila swarms he had accumulated. The fly, a male, was crossed with a red-eyed female. All of the 1,237 offspring had red eyes. When the offspring were interbred, the young were red or white in Mendel's 3:1 ratio, as we would now expect. In addition, the white-eyed flies were all males. Not only were white eyes a recessive trait but they were gender-linked. The white-eyed trait was finally traced to the bent member of the XY chromosome pair.

Tracing the white-eyed trait to a definite chromosome, the male Y chromosome, was a triumph for Darwin and Mendel and the beginning of a long series of studies into the positional structure of gene sequences. It would be found that it is not the chemical composition alone, but also the number, sequence, and position of the atoms and molecules involved, that controls the processes of replication, development, and maintenance within an organism.

Bibliography

Alberts, B., Bray, Johnson, Lewis, Raff, Roberts, and Walter, 1998, *Essential Cell Biology*

Atkins, P. W., 1991, *Atoms, Electrons, and Change*

Burchfield, J. D., 1975, *Lord Kelvin and the Age of the Earth*

Darwin, C., 1859, *The Origin of Species*

Dawkins, R., 1976, *The Selfish Gene*

Dawkins, R., 1986, *The Blind Watchmaker*

Dennett, D. C., 1995, *Darwin's Dangerous Idea*

Dunbar, C. O., 1960, *Historical Geology*, 2nd Ed.

Edey, M.A. and Johnson D.C., 1989, *Blueprints*

Eiseley, L., 1958, *Darwin's Century*

Encyclopedia Britannica, 1989, a general aid

Futuyma, D. J., 1986, *Evolutionary Biology*

Gould, S.J., 1996, *Full House*

Grant B.R. and P.R., 1989, *Evolutionary Dynamics of a Natural Population*

Johnson, P. E., 1991, *Darwin on Trial*

Mayr, E., 1991, *One Long Argument*

Mayr, E., 2001, *What Evolution Is*

Margulis, L. and Sagan, D., 1997, *Microcosmos*

Margulis, L. and Sagan, D., 1995, *What Is Life?*

Margulis, L. and Sagan, D., 1997, *What Is Sex?*

Nobile, P. and Deedy, J. eds., 1972, *The Complete Ecology Fact Book*

Nordenskiold, E., 1935, *The History of Biology*

Johnson, P. E., 1991, *Darwin on Trial*

Sapp, J., 1994, *Evolution by Association*

Schopf, J. W., 1999, *Cradle of Life*

Strickberger, M. W., 2000, *Evolution*

Tassy, P., 1991, *The Message of Fossils*

Villee C. A. and Dethier, V.G., 1971, *Biological Principles and Processes*

West, G. H., 1938, *Charles Darwin, a Portrait*

Chapter 10. Individualism and Evolution

Evolutionary science is largely concerned with the investigation of species and populations of individuals. A major biological question is how variations among individuals can lead to new species. To achieve an adequate understanding of the processes involved, knowledge of every species, present and extinct, must be rationalized into a coherent, overarching explanation. Because of the complexity of the data, answers were slow emerging. It is remarkable that Darwin's theories of evolution, developed during the eighteen hundreds, met every test while defeating every attack. The attacks only demonstrated that no other theory or logical explanation approaches the explanatory power of Darwinism. The two Darwinian theories, that all life descended with modifications from earlier living forms and that natural selection is the main operative process that facilitates the change of individuals into new species have become more firmly established each year. That is not to say the future will not bring change and extensions to the theory as new facts emerge.

The Fossil Evidence

Fossils of varying ages are broadly scattered over the surface of the earth proving that earlier living forms existed and changed; they evolved. Museums are awash with fossils. These are objects we can observe directly. Everyone can see fossils, feel them, and reason about their origin. The only logical explanation, to date, is that fossils represent the remains of long dead individuals, many from extinct species.

The shape, size, and age of fossils often show similarities indicating they were from individuals of a common lineage that has changed over time. Fossils of the leg and foot of Equus, the horse, are often cited as examples of how species can change. Early fossils of the line that led to Equus show a steady sequence in the number of toes from four, to three, to one bone (the hoof) in the modern horse. Many other features of the horse, such as the teeth, jaw, and skull, also show obviously related and continuous changes from earlier to more recent specimens. All species for which we have found sufficient fossils show similar relationships.

Natural Selection

The role of natural selection in the overall theory of evolution is sometimes difficult to unravel because it consists of two major processes that are complex, unfamiliar, and interactive: variation and selection. What varies; what is selected?

It is both the individual creature and its surrounding environment that vary. Changes in an individual's internal environment, for instance, variations in chemical concentration or genetic changes during the processes of meiosis, crossover, and reproduction, lead to different individual characteristics. Because of these changes, living creatures are always unique. Even identical twins have at least some internal variations due to errors or changes during replication. There are no individuals identical to you; there never has been and there never will be. Random chemical and structural changes during the replication of DNA can cause variations that are beneficial, lethal or anything in between. If body cells become defective they can sometimes be replaced, but if the reproductive cells are involved, the defects are passed on to the young. This is not true of single-celled creatures such as bacteria; they reproduce by simple cell division and pass all changes on to the next generation. In function, a unicellular creature is also a reproductive cell.

Another part of variation, the external environment, includes such variables as climate, predators, competitors, chance, and food supply. If the environment remains relatively constant, well-adapted species will tend to adapt even better over time. If the environment is constant and also benign a wide variety of individuals can survive. In this case, when the environment becomes harsher, many variations will be eliminated, but because of the differences between individuals, a few are likely to survive. These few will then multiply to fill the harsher environment and save the species from extinction. A narrowly specialized species is in greater danger of extinction than a more varied one. Poorly adapted individuals in one environment can sometimes become the well adapted in another.

The impact of natural selection on individuals is often described as violent, savage, vicious, dreadful, and pointless. But is anything more violent and pointless than the death that awaits all individuals? Is death pointless? Without death there would be no selection of individuals with the ability to survive in more complex and changing environments. The first repeating, self-replicating and stable molecules or creatures would have quickly sequestered all the available carbon and phosphorous atoms necessary for life with nothing left for other replicating creatures. Variation and selection led to the great variety of species on earth today. The negative phrases that are heaped on natural selection are meant to discredit the scientific concept, not to describe it. They also do not disprove it.

Yes, natural selection is in human terms violent, savage, vicious and dreadful, but it is not pointless. Natural selection has outcomes; we are one of them. Are we pointless? Not unless we make ourselves so.

Natural selection is also described as an active process in which nature selects some and kills all others. Such is not the case; the environment independently follows its changing nature; it is the individual, attempting to live in an environment dictated by chance, who succeeds or fails. In any environment there are innumerable individuals attempting to live and reproduce. The environment gives no choice or exception or explanation to anyone. It is simply there; it actively selects nothing. It has no end purpose; it is we who see a purpose, after the event and for our human needs and understanding.

In fact, some of the factors involved in natural selection have little or no violence associated with them. The evolutionary process requires that individuals both live and reproduce. Half of this requirement, the ability to reproduce, is seldom violent. We all have seen pictures of two mountain sheep running full tilt at each other and butting horns to curry the favor of a female standing discreetly off to the side. Invariably, one of the two will decide he has had enough and saunter off. Such encounters seem violent but are seldom deadly. Similar behavior is seen in the sexual competition among most animals and birds. The bonds between male and female, mother and child, and male and herd are normally nonviolent, requiring considerable cooperation among individuals.

Food is the most critical need of life. The sparrows spend their whole day searching for seeds but seldom will you see them fighting over one. If you put down a tray of mash for a flock of hungry chickens, they will all rush to the tray, even jumping on each other to get to the food. But this is nothing more violent than a crowd of hungry guests pushing to get to a splendidly-arrayed banquet table. A severe scarcity of food leads to a slow death for any population, bird or man. Those who are good at searching for food and can live on little have the best chance of survival. Recall the Galapagos finches; they periodically starved almost to the point of extinction because of a lack of rain. During droughts, only the tough-shelled cactus plants produced seeds; consequently, finches with the strongest bills survived.

There are additional factors involved in natural selection such as catastrophic climatic changes, chance accidents, instability of internal conditions maintained by the body (for example, temperature), and predators. The activity of predators is the main source of the idea that natural selection is a violent, vicious, and savage process.

The Gifts of Natural Selection

From a human point of view, existence is the primary good and life the greatest gift. But there have been many lesser ones along the way. For example, in the early universe of life, the earth had a chemically reducing atmosphere that was mainly water, hydrogen, sulfur oxides, hydrogen sulfide, and carbon oxides. Early life forms used these reducing materials as an energy source for their metabolic needs. About 3.6 BYA, photosynthetic bacteria evolved that could convert water, sunlight, and carbon dioxide into a simple organic compound — glucose and oxygen. The increasing levels of oxygen resulted in the large banded iron oxide strata now common on Earth. But more important to us was the emergence of life forms that could use oxygen metabolism with its greater energy value needed for complex movement. Photosynthetic bacteria increased the oxygen content of the atmosphere from less than one percent to 21 percent. They are still, together with green plants, actively supporting us with oxygen. For at least three billion years they have been giving us this gift.

A third gift of natural selection was the emergence of bisexuality about 1.5 billion years ago. As described elsewhere, bisexuality led to the emotions of love and compassion, the family, clan, and society. And with humans, consciousness began.

But there are disadvantages as well as benefits to the process of natural selection. The primary evil, of course, is death. But without death there could be no natural selection or evolution; we would not exist. Is it a disadvantage that we exist? No, we believe existence is the basic good. To repeat, without universal death the first replicating molecules would soon incorporate all the relatively rare elements necessary for life. The result would be that the replicating process would cease for lack of raw materials, and evolutionary change would end. Fortunately, ultra violet radiation, cosmic rays, high temperatures and so forth readily break up the carbon-based organic molecules we are made of. Unfortunately, this relative instability also leads to death. Because of these disruptive processes, there is equilibrium between the rates of replication and death. Between these alternating processes, carbon and other elements again become available and new molecular structures can form, which reintroduces the possibility of natural selection and evolutionary change.

Species

All individuals die, species never die; they become extinct. Between the two — death and extinction — there is an enormous difference. Only after every individual dies do species cease to exist. There is no pain or terror, only a conclusion, an end, a break in the chain of being. In contrast, individuals often die in the extremes of pain and terror. But even individuals seldom take much note of the death of one of their own species. Nature knows that excessive concern for the dead can endanger the living. Occasionally a female polar bear will stay with a cub that has died for awhile, but soon she moves on to care for a second cub or follow its own needs and instincts. As far as we know, humans are almost the only creatures to feel sorrow, dread, and emptiness at the loss of another.

Natural selection also imposes limitations on the ability of an individual to adapt to changing circumstances. DNA controls abilities and these molecular structures are fixed for each individual. In a sudden drastic change of environment an individual cannot adapt beyond the capabilities of its DNA. Most species cannot reason beyond their instincts and perish in the face of severe changes. We can put on heavier clothing in the face of extreme cold, but a duck or hummingbird cannot.

When individuals have adapted to a new environment, an additional sudden change can cancel the advantage gained. Often the most highly adapted species box themselves into an evolutionary dead end from which they cannot retreat. It is a race; a species must adapt at least as fast as the environment changes but not too much more. Adaptation in this context means to have undergone natural selection successfully.

There is a mixed advantage to our lack of direct control over the abilities and drives that make and sustain our lives. Our sensing organs, for instance, were given to us unbidden and we could not improve their sensitivity or range until recently. But it was probably fortunate we had no control over the development and fixation of DNA. Such skills were totally beyond our talents or knowledge. In addition, reason has seldom gained ascendancy over instincts. Our inherited drives to procreate excessively and exploit the earth's resources for the sole benefit of man, are examples. So there is a mixed benefit to our lack of control over the DNA abilities of our bodies. However, we are now on the verge of micro molecular control over the DNA that dictates many of our functions. Whether we will be able to use this new ability for the benefit of all is an open question.

How can a species be defined? For animals and other bisexual creatures, the usual criterion is whether two individuals can interbreed successfully. The donkey and the horse can interbreed, for instance, but their offspring, a mule, is always sterile. Most interbreeding between species cannot get as far as that. Many species have varieties or subspecies that can interbreed relatively easily. So there is often confusion as to what is, or is not, a true species. With the developments of molecular biology we have a more rigorous method of definition. The genetic structure of two species can be directly compared for differences. But even with this refinement, arbitrary decisions will still have to be made. We should note that more than 95 percent of human and chimpanzee genes are identical and 99.5 percent of Neanderthal genes are identical with ours.

The principle that a species is defined by the ability of individuals to interbreed is the clearest we have for animals. Unfortunately, nature seldom gives a single answer to a single question. The definition might hold for species like ours, which are bisexual and multicellular, but not for those numberless species such as protozoa, bacteria, yeast and algae that are usually unicellular. They reproduce by division into two identical daughter cells. There is no trading or matching of genetic material. Unicellular creatures preceded the multicellular by about 2 billion years and still provide us with such basic needs as atmospheric gas balances, energy conversion, and decomposition processes.

Individuals

The question of species is important, but we are not species; we are individuals. The role and fate of species and those of individuals can hardly be more different. Species move along serenely, increasing or decreasing, changing or static; it matters little. Species are man-made abstractions containing no atoms, no consciousness, no senses, no pain, joy, or sorrow. When faced with extinction, they do not recoil in horror or cry out in pain. None are even sad, except perhaps onlookers like us, who can look into the future and see.

In addition to individuals there are families, clans and societies before we arrive at species. Individuals, families and to some extent clans are blood relatives. In societies the blood relationship becomes ever more dilute.

It is we, the individuals, who suffer and die under the hazards of our surroundings. We alone carry the whole burden of natural selection. Our struggle occurs at each existing moment as we blindly use our genetic endowment to live and reproduce within an environment we never chose.

Species can evolve into new species; we cannot change into new individuals except in a metaphorical sense. Our genes hold us to a definite physical structure that cannot be changed into another. Some are led to think we are only vehicles designed by genes to achieve their own immortality (Dawkins, 1986). This is a somewhat teleological view that hardly avoids a claim that genes somehow plan their own evolution. But genes cannot freely select their chemical makeup. What genes do is the fixed result of individual variations and selection — the processes of natural selection. These processes depend on chance. For instance, a single high-energy photon that happens along can disrupt the molecular structure of a gene and consequently cause the death or mutilation of a fertilized egg cell, an embryo, a fetus, and even an individual.

We are so far from our chemical origins that we hardly grasp the magnitude of change we represent. But with the continuing discoveries of molecular biology and paleontology we see more clearly where we have been, and through what we have come. We conscious individuals must decide where we are going. Can we ask then: who created; who has been created? Unbidden, our ancestors replicated themselves and created us. In a certain sense each living individual creates itself and its future. From within the confines of our DNA instructions, we must win or lose with weapons we have never used, against foes we have never seen.

How can we thank those endless, unknown generations that made us? They never looked for thanks, as we shall not. They tried to live with what they had, as did their peers who often died in the attempt. We must honor them all. Each stands as a colossus on the fields of trial and success or failure that once surrounded them and now surround us.

Each moment we live with the knowledge of our genes as a shield. Forged in tests through billions of generations, genes throw up our defenses and escort us on to another day. We must learn their language and do their bidding. Still, how many have followed DNA and failed anyway? Many, most, all have ultimately failed; for the environment will not stay, the variations will not cease. In a universe of change, to be something, we must change into something else.

The Individual and Reproduction

Although the earliest stages of life are well beyond our present state of knowledge, we can lay out a simple scenario of how the event could have unfolded. The first forms of life were most likely simple molecular compounds that possessed the major characteristics of the evolutionary process — the ability to grow, endure and replicate. Later, these replicates surrounded themselves with a wall. Within the wall, materials were accumulated and processes selected without interference. During this era, reproduction occurred by molecular replication (mitosis) and division of each cell. Bacteria, algae, and protozoa are all unicellular creatures that reproduce by dividing into two identical individuals. Some bacteria have alternate methods of transferring DNA without cell division. E. coli can transfer DNA directly through the cell wall to another member of its species. Other bacteria can absorb free-floating or naked DNA from their fluid surroundings (Alberts, et al, 1998). But normally DNA is maintained intact within the cell.

With the emergence of multicellular life came an elaboration of complex and sensitive processes that created cells with new capabilities and responsibilities. Some developed the ability to move and swam in the early seas encountering nutrients on the way. Others developed chlorophyll and were able to convert radiation from the sun into forms of energy useful to the organism. Still others were sensitive to light variations and the difference between night and day became detectable. This ability, through the long and tortured path of evolutionary change, eventually became the eye. The processes of single cell maintenance and replication were carefully honed for about two billion years before the multicellular revolution took hold.

The unicellular form of reproduction continued exclusively until about 1.5 billion years ago. Then somewhere and somehow an individual divided into two slightly different beings, each unable to reproduce until they combined their DNA with the other — the defining sexual act. Some have marshaled evidence that the event occurred when one individual absorbed an individual of a different species, perhaps a bacterium (Margulis, and Sagan, 1995). The two individuals lived together symbiotically, each contributing its special talents to the common advantage.

However the event occurred, it was a happening of momentous importance. Until this place in evolutionary time, individuals were independent and complete. Sometimes they gathered into colonies, some traded DNA by a variety of methods, but usually they lived and reproduced alone. The bisexual species had a strange new burden to carry. They consisted of two kinds of members — male and female. They could not reproduce alone; they were incomplete. One of the first requirements and definitions of life is the ability to reproduce and these creatures could only reproduce with the help of another individual. Nevertheless, this was the path that new creatures like plants and animals would follow. This is our fate; we must join with another, or be incomplete and finally extinguished. Here, at this juncture, are the beginnings of love, compassion, family, clan, and society.

Life was setting out on a different path. For a long time there was little notice in the fossil record of any great change. But below the level of our vision there must have been a movement toward greater variation and complexity. Seemingly in a brief moment about 550 million years ago, multicellular life flowered in the Cambrian explosion. New kingdoms of life — plants, animals, and fungi appeared for the first time. This elaboration of new creatures is most perfectly preserved in the fossils of the Burgess shale, 8000 feet above sea level in the Canadian Rockies (Gould, 1989).

In the beginning the two individuals involved were not significantly different. Perhaps the unicellular creature could no longer divide into two identical daughter cells; however, by combining DNA with another individual, division could occur. Whatever happened, the end result was the same. Individuals could reproduce only if they shared DNA with another; they had become bisexual. The two forms of reproduction, unisexual and bisexual, have coexisted for a long time. To this day some species are able to reproduce by either method.

Advantages of Bisexuality

For bisexuality to emerge, it must have had advantages. The disadvantages are obvious. Each individual must find another, similar but somewhat different kind. There must be mutual recognition, attraction, and acceptance. There are also complex problems associated with reliably transferring DNA from one individual to another and finally the combined DNA must develop into functioning adults. All of this required a long period of mutation, selection and stabilization to achieve a method of reproduction. The time and energy needed were undoubtedly enormous. To us, time and energy are important but evolution has no time constraints because it has no plan, no goal; it is going nowhere by intention.

At least three advantages made the effort worthwhile. First, the new species could become larger and more specialized. Second, the redundancy of two sets of chromosomes allowed for repair of damage and replication errors in the DNA. And third, the interchange of segments of a chromosome pair (during meiosis) fostered greater variation in members of the species.

Most unicellular creatures are roughly spherical. Spheres increase in volume as the cube of the radius whereas their surface area grows only as the square of the radius. This geometric difference means the volume of a cell grows much faster, in a progression of 1, 8, 27, 64..., compared to the surface area which increases as 1, 4, 9, 16, etc, times the radius. This increasing disparity between volume and surface makes cell size self-limiting. The surface membrane of the cell must transfer all food and building materials into the interior. A limit to the rapid diffusion of ions and molecules through the cell wall is soon reached; the cell becomes inefficient and stops increasing in volume.

Some modification of these limitations can be achieved by assuming a thread or sheet like form but the advantages of a multicellular structure are overwhelming. Separate cells can concentrate on separate functions. To achieve a multicellular state, the DNA had to take on a complex new set of capabilities. By accumulating separate cells for special tasks such as motion, energy conversion, internal communication, and sensing, vast new abilities became available to a multicellular individual. During this life era, DNA was enclosed in a sac of its own — the cellular nucleus.

The Emergence of Choice

Over the last four billion years, life has emerged in a world of chance — chance individual variations and chance environments. These were thrust upon us. But now we humans are faced with chance and also choice — many choices. There is no longer the necessity to follow all the dictates of natural selection. As conscious beings we can sometimes see the long-term consequences of our actions and as individuals we can act on such knowledge. We can make wise choices but first we must defeat ignorance.

We are like cells in an organism. Every living cell has chromosomes that contain its life instructions. We can be like that DNA; we can obtain and act on the knowledge needed to help our world continue in peace and prosperity. But knowledge alone has no power since it cannot force acceptance or action. And, unfortunately, the necessary knowledge is in the language of science, often thought tedious, opaque, elitist, and not very compelling. Science gives no unending flights of poetic fancy or fictional entertainment. It makes a choice for sharp precision and routines that often seem dull. But the overall vistas that spring from science are more awesome and chilling than anything philosophy can even imagine. The sheer quantity of scientific knowledge can discourage us from the effort to understand. Indeed, excursions along the outer reaches of science are often beyond understanding. They are the outposts of specialists. But hopefully, through complete education for all we will have the means, the basic scientific knowledge, required to plan and act for a benign and livable world. We will then be able to make wise choices for the future. The role of chance can be somewhat reduced and circumscribed. But all of this can come about only through individuals who are well educated and participate cooperatively in the functions of society.

Purpose

Purpose arises with the beginnings of life in the universe about 3.9 billion years ago. Living things have such needs as food, air, and water. They interchange materials with their environment. With life, purposeful action first comes into existence. We decide what purpose is, what has it, and what does not. However, our language is sometimes ambiguous, causing us to imply a purpose where there is none. For instance, a statement such as: "There is a species of shrimp that swims about in the moray eel's mouth cleaning its teeth," is not quite accurate. The shrimp are eating scraps in the eel's mouth but their purpose is to find food; they have no interest in cleaning teeth. The example is trivial, but it shows how we often cite a purpose that is wrong or nonexistent.

A search for purpose seems to be a special characteristic of humans. It is often expressed as a plan, goal, or reason for being. Or it is felt as a desire for importance or being needed. All living creatures have the constant need and purpose to acquire nutrients from the surroundings in order to sustain life. Our impulse toward purpose is part of this life-sustaining requirement.

A sense of purpose depends on your viewpoint. The wild horse searches for grass and water. His purpose is to live or more exactly to satisfy the immediate cravings of his stomach. The grass does not share this sense of purpose; its purpose is to live by gathering water, light, and nutrients to grow root, leaf, and seed. The water, of course, has no purpose in providing drink for horses, plants and men as it moves calmly down to the sea. The grass might look on the horse as serving a good purpose by keeping the grassland cut short so that shading trees and shrubs cannot grow. A man looking at this scene would focus on the horse. The purpose of the grass, he would decide, is to support the horse; the purpose of the horse would be to serve by carrying him long distances, or pull a plow, or provide food or something to trade. He would take care of the horse; Arab nomads, for instance, were known to take horses into their tents during sandstorms. Nomads, in general, give dedicated care to their livestock. When the grass has served its purpose and grows sparse, man and horse see purpose in moving on to new grazing grounds.

Now, consider some objects from the inanimate physical world — the sea and the sand. Under the urging of the tide, the sea lashes toward shore with increasing rage and fury. The waves reach out like huge hands with angry fingers ready to grasp the sand and carry it off. The sand is engulfed, swirling up and over, carried up the beach, and slowly released to settle back into a new tranquility to await the next assault.

What is happening here? Is the sea's purpose to wear away the shore, to carry off the sand, to spend its energy? Is the sand determined to stay where it is? No, there are no aims, goals, or purposes here. The sea has no rage or angry fingers, the sand no tranquility, no memory of previous assaults or anticipation of another. Sea and sand are interacting under the laws of gravity and science. The end result is that the shoreline is slowly changed; the sand is ground into smaller and smaller particles that are entrained and carried off by the sea, but in all this there is no sea goal or sand goal achieved. In like manner other interactions between inanimate objects are without purpose whereas the activities of living creatures are always full of purpose — their own.

In the physical universe there is no purpose before existence; change precedes purpose. After the change we assign the plan and the purpose. In the human world we first aim to achieve a purpose — purpose precedes change. Our purpose is to go to the store because we are hungry; therefore, we plan to take along some money for groceries. The universe has no plan. It is simply there. It is made of material particles and photons obeying their random, probable and group natures as some climb the ladder from aimlessness to life. Only with the emergence of life do we find purposes. All change has a result and the result is often incorrectly called purpose.

The inanimate world follows a general path toward decreasing free energy and increasing entropy. It becomes dilute, chaotic, and unorganized. It seems to be fleeing from its fiery origins in the Big Bang toward a featureless haze of nothing.

We create purpose; we find it everywhere. When Quakers first saw the land at the confluence of the Schuylkill and Delaware rivers in 1681, they immediately saw its "purpose." Here was a fine natural harbor for a town; it had good fresh water for a port and for drinking; the forests could provide timber to build houses and ships; there were plenty of fish and wildlife; the surrounding land was fertile. Everywhere they found features suitable for their purpose as planners of a new town in a new world, later to be called Philadelphia.

Complex Purposes

The human eye is often cited as a complex structure that could never be produced by the processes of natural selection and evolution. The argument is that anything simpler than an eye, such as half an eye, would be utterly useless; therefore, the human eye must have been created instantaneously and whole. Such claims are false; they fly in the face of all the facts we know from observation and reason.

The basic error in such reasoning is thinking that nature can produce complex structures in a single step. Dust clouds in space do not condense into suns and planets in a moment nor do dinosaurs become birds in an instant. Micro changes, over long periods of time as compared to a human lifetime, characterize the processes of evolution. Sudden large changes in DNA often lead to the early death of individuals. As we have seen, DNA is a very complex molecule both in structure and function; any variation in as little as a single atom or nucleic acid base can lead to interactive changes in other parts of the DNA helix. All such changes must then be tested and proved functional by natural selection. Excessively large changes increase the probability of death. Large mutations, once thought by de Vries and others to be the engine of species creation, instead are often found to be lethal.

Eyes have evolved in about fifty different invertebrate groups in at least nine distinct basic patterns. These include designs based on pinholes, lens, reflectors and compound structures (Dawkins, 1995). Almost all vertebrates have eyes. Some species have eyes that degraded over time due to decreasing use; these include cave dwelling fish, bats, and moles. In comparison, owls and other night creatures have eyes with extreme light sensitivity. Fossils of many species dating back over the last five hundred million years show clearly from the eye socket configurations, for example, that eyes evolved over a very long time.

Light sensing began with simple organic molecules that change into a different form when exposed to light. Many organic compounds are changed into excited (or reactive) states when they absorb a photon of light. The excited state is normally unstable and soon returns to the stable or ground state. This change between two chemical states produces electron transfers that could send signals to a central nervous system indicating light intensities have changed. Another important photon-molecular interaction is the energy conversion process carried out by chlorophyll when it absorbs visible light to make glucose and oxygen from carbon dioxide, water, and photons. Life's use of this photon/molecular interaction dates back more than three billion years when photosynthetic bacteria began using it and also produced the oxygen we now breathe and the fuel we burn.

In the human eye, the light sensing molecule is called rhodopsin, which is a molecular combination of retinal, derived from vitamin A, and opsin, a protein. When rhodopsin absorbs a photon of light, the retinal portion (called cis-retinal) is changed into its isomer (trans-retinal) and the opsin portion is released. This cis/trans exchange, carried out in several steps, triggers an electrical nerve impulse. Through the action of two enzymes trans-retinal is converted back into cis-retinal, which reunites with the opsin portion of the original molecule. The process is then repeated as long as rhodopsin is exposed to light (Lehninger, 1986)

The primitive eye, beginning as a spot of rhodopsin — like organic molecules, was so valuable for survival that it rapidly evolved into more elaborate and sensitive structures. Knowledge of the evolution of eyes can be traced by microscopically examining eye structure in adults of many species. Bird egg embryos show a similar sequence of structures as they develop before hatching. The evolutionary sequence is roughly: first, a flat section of tissue containing the light sensitive molecules forms, second, a concave shape, third, a cup, fourth, a jug with a narrow neck, fifth, lens formation within the jug neck which finally becomes the focusing eye (Strickberger, 2000). The eye has evolved over long periods of time in a seemingly endless and seamless series of steps.

Humans in Space

Sometimes we feel like Masters of the Universe, but perhaps the wine saga of Saccharo E can tell us something. In many ways, the Earth appears to be a closed system too, but fortunately we get a supply of new energy from the sun each day. Unfortunately, we are rapidly depleting our supplies of coal, gas, and oil, which are the fossilized photon energy of the past. As we burn these fuels ever more rapidly, we are poisoning and overheating earth's atmosphere. Our experience with atomic energy as an energy source has been mixed at best because of the dangers of radioactive contamination, explosion, and waste. Fossilized fuels and atomic energy, are the only plentiful energy sources within our present technological capabilities. Atomic fusion has been in the research and development stage for decades and shows a decreasing probability of success although recently there has been new interest in the subject. Wind power and solar cells can provide energy but only for relatively small applications compared to our present level of energy use. And as for automobile fuel cells running on hydrogen from water, it requires more energy to split hydrogen from water than you would get in return from the reaction of hydrogen with oxygen to power a fuel cell. Hydrogen from water will therefore, cost and pollute more than fossil energy does today and that is without considering the enormous cost of distributing millions of pounds of explosive compressed gas. These last statements are based on the immutable laws of thermodynamics, which have never yet been violated to our knowledge.

Declining sources of energy and food are not the only problems for us. As with Saccharo E there is also the specter of an exponentially expanding population. Conditions at the moment are benign for us; we truly are the masters of our universe. But most other species are withering to the point of extinction under pressure from our activities.

We are facing all of Saccharo E's problems; decreasing food supply, increasing wastes and population. We might send a few humans off in spacecraft to look for other worlds much as a few Saccharo E escaped by chance in bubbles of carbon dioxide. But that is not a possible solution for us. First, where would we go? We know of no planet that could support humans. Second, the energy required to put the world's population into space is well beyond foreseeable human capability. Someday we will probably be able to send a few humans off to populate another planet but would the great majority of us stay behind willingly? The goad of mass starvation would lead to a climate of inhumanity beyond imagining. When societal restraints collapse we have often revealed ourselves as beasts no less vicious than the wolf pack, the lion pride or the piranha school. At least these three do not often kill their own.

Where would we go? Science fiction writers amuse themselves by blasting off to a nether land in search of other worlds. They mix in enough jargon to bring the confirming breath of science to their stories. Unfortunately, confirming evidence is lacking at the present. The maintenance requirements for emergent life in the universe are immense and the overwhelming role of chance is clear. Simple forms of life such as bacteria or molecular complexes that can reproduce probably exist in other places in the universe but the possibility of more complex intelligent life is remote. It is not only that the requirements are many, but also large evolutionary events such as extinctions must occur in a definite sequence and at certain time intervals. All of this brings the operation of chance into a dominant role.

By searching for the conditions that favor the genesis and elaboration of life on Earth, we can estimate the probable requirements for life elsewhere. Some of the goals that must be met are: significant amounts of heavy elements, relatively constant temperature, and orbits, meteors, and comets of the proper number and size (Ward and Brownlee, 2000). The early universe had no elements heavier than hydrogen and helium. Heavier elements became available as dying stars entered their brilliant supernovae phase about two billion years after the Big Bang. Elements such as carbon, hydrogen, nitrogen, oxygen and phosphorus are the basic materials of life; without them life could not emerge. Also heavy metals such as iron, nickel and cobalt form the partially molten metal core that generates earth's magnetic field. The field provides a magnetic shield against the ionized particles coming from the sun in the "solar wind." Without the shield, high energy ionized particles would quickly sterilize our planet except for simple forms of life living underground or in the deep seas.

Simple life forms have recently been found living under hostile conditions around volcanic vents on the ocean floor. These hostile conditions include increased pressure, temperature, and lack of an oxygen atmosphere. Such discoveries show that the conditions for life are much wider than formerly thought. Another positive development is that planets circling other stars within our Milky Way are being discovered almost weekly. With newer telescopic equipment and techniques we can expect increasing numbers of planets to be found.

Whether we could inhabit the newly discovered planets is questionable. They are either too small or too large, too near or too far from their sun to provide the correct amount of energy for a suitable surface temperature. From spectral analysis we find most of them are made of gas not rocks, and so on. In time we will certainly find a planet similar to earth that could support our kind of life. Whether the planet will be near enough to reach is another matter.

For our kind of life to evolve a plentiful supply of liquid water must be available. Liquid water freezes below 0 degrees and vaporizes above 100 degrees centigrade so the planet temperature must be stable within those temperature limits. Also, DNA the basis of our inherited characteristics and replication processes, is only stable up to about 90 degrees centigrade; above this temperature the double helix spontaneously unzips or separates into two strands. In addition the temperature must be stable for extremely long periods of time. At least parts of our planet have benefited from temperatures between +/−100 degrees centigrade for 3.5 billion years or more.

Now the duel with chance becomes obvious. The big bang occurred about 13.7 billion years ago. The earth formed about 4.5 billion years ago. After about another billion years, replicating, single-celled creatures began evolving on earth. It took about 2 billion years more before bisexual, multicellular forms appeared in the form of plants and animals. Why did it take so long? Will it always take that long?

Alone in the Throes of Change

Because we are social creatures we seldom realize that, in a profound sense, we are alone. Surrounded by family, friends, and society we carry on our daily lives in a whirl of activity. But below this level we breathe, avoid danger, find food, and generally care for ourselves alone. We cannot breathe, eat, feel pain, or die for another. No other individual's heart can pump your blood. It is a penalty we pay for our separate and unique natures. It is the price of change. Each individual has a somewhat different plan for survival. If our DNA has a tendency toward cancer, we alone must suffer. Even with the support of medical science, society, and family, we alone make the final passage to death. The individual bears the whole burden of natural selection. We are both the vehicle that carries evolution onward and the worker who acts and sacrifices in the service of natural selection.

It is the particulate nature of things that separates us as unique individuals. Whether particles, atoms, or men, we stand naked in existence, alone. Particles and photons dance their jig of creation, change and annihilation; atoms form shared bonds of harmony; and we reach out hopeful arms toward mate, family and society. Yet in essence all these are by nature and purpose, alone. It is characteristic of things and creatures to fold themselves into knots of exclusion, fearful of the world. We are individuals, alone in a continuum of life and must act largely as nature requires.

The creation of life as discrete individuals is a necessary prelude to evolution. Since we are accountable primarily to ourselves, we do not endanger our kin or species if we falter or die; there are others to carry on. If we were so closely interconnected that the failure of one would weaken or bring destruction on all, evolution would end.

Most unicellular creatures are even more isolated. When they reproduce by binary division, one becomes two, but both are completely independent and alone. Each has the total responsibility to live, to acquire nutrients, build DNA and proteins, avoid danger and reproduce. There is no family to help with the problems of life or society to expand and extend the benefits of shared activity.

From the earliest times, life built cell walls to exclude the environment, resulting in greater control over basic chemical processes. This was a fundamental change that introduced individuals into the universe and allowed natural selection to operate. But the greatest act of exclusion was the separation of an individual into two parts, male and female. No other happening has so dominated our individual lives. We each search for that lost half, the source of love, family and society. A bisexual creature must find and unite with the other or face total extinction. Families and societies are built on this overpowering individual obsession. And all of this is a result of natural processes favoring variation and selection. From this perspective, natural selection becomes a gift most glorious.

The process of natural selection gives us the gift of everything — life itself. The complexity of life is beyond our comprehension. Consider: our bodies contain a trillion cells; each cell is teaming with trillions of atoms in a swirl of chemical reaction as it builds genes and proteins; each genetic trait can undergo ten thousand million mutations during a lifetime. The numbers alone defeat our efforts at understanding. Who would have taken a wager 13.7 billion years ago that creatures like us would someday walk the planet earth while it orbited a common star? We are truly creatures never meant to be.

The Natural Selector

We are living in a strange interlude. Our mastery of nature has become so complete we no longer incur the full force of natural selection. We are taking the first steps into an era of human selection. Under the easy environment we have created for ourselves we breed and monopolize space and materials essentially unrestrained by natural selection. It has been a partial triumph of intelligence over the physical forces of nature but more like a passing dream.

Defective genes that were previously eliminated by natural selection are inundating our genetic constitution — our genome. We have become the natural selectors before we have the skill or will to provide a benign and rational substitute for nature's processes. We are domesticating the entire planet. Vaccines and antibiotics have almost eliminated infant infectious diseases in the developed countries. Improved crops have eliminated famine, again in the developed countries. There is evidence of improvement in the quality of air and water in some areas and the rate of population growth may be slowing. But wherever there is political or economic turmoil there is war, disease, and famine. We live on the edge of disasters of our own making. These are a different kind from those of our ancestors. Only a few sense or understand these dangers.

We have emerged as the Natural Selector. Our collective behavior controls the fate of all species. Growing more dominant each day, we feign disbelief to cloak our greed, ignorance, and self-interest. We have elected for our companions a few dozen "useful" species: wheat, rice, pines, oaks, fish, cattle, etc. All the rest are tentative, on the edge, unknown, or abandoned. Their extinction events are left to the momentary flutter of public opinion and enthusiasm.

Because of our empty concerns, environmental catastrophes are inevitable. How many disasters can we survive? Hopefully they will begin and spread slowly so we can adapt quickly enough. Thorough education, consensus, and action are our only hope short of becoming robots, or masters and slaves of the absolutes that crowd in upon us from dictator states and rampant expansion. Yet looking to our history, have we ever agreed on anything without strife and bloodshed?

Bibliography

Dawkins, 1995, *River Out of Eden*

Alberts, et al, 1998, *Essential Cell Biology*

Gould, 1989, *The Burgess Shale*

Lehninger, 1986, *Principles of Biochemistry*

Margulis and Sagan, 1995, *What Is Life?*

Strickberger and Monroe, 2000, *Evolution*

Ward and Brownlee, 2000, *Rare Earth*

Whitrow, G. J., 1988, *Time in History*

Chapter 11. The Double Helix

In the early 1800s, organic chemistry was considered the chemistry of materials produced only by animal or plant life. But in 1828, Friedrich Wohler synthesized urea, a simple organic compound, from three inorganic compounds: lead cyanate, ammonia, and water. Wohler's discovery opened the way to the synthesis of millions of organic compounds, many times the number of known inorganic compounds.

The first important step toward an understanding of the inheritance of traits in living creatures was made by the Swiss biochemist, Friedrich Miescher, in 1869. He found a way to separate cell nuclei from cells by using a digestive enzyme. This allowed him to do a chemical analysis of chromosomes without interference from other materials in the cell. He called the white powder obtained, nuclein. The modern formula for nuclein is $C_{29}H_{35}N_{11}O_{18}P_3$; the modern name is nucleic acid. It has the basic chemical elements of DNA (deoxyribonucleic acid). Although Miescher's original formula was a little too high in hydrogen and oxygen and a little too low in nitrogen, it was remarkably accurate for 1869.

By the middle 1800s many chemists were working on the synthesis of organic proteins, the major constituents of living tissue. Although Miescher's material was an acid, many chemists believed it looked suspiciously like a protein needing only a little more purification. Much like Mendel's work of 1865, Miescher's of 1869 was largely ignored. Here were the most important discoveries of the 1860s, yet no one had the slightest idea what they were or what they meant because the scientific methods necessary to unravel such mysteries had not yet been developed.

X-Rays Crystal Structures — Von Laue and Bragg

The most important missing method was how to measure the position and spacing between atoms in a crystal or molecule. It was not until 1912 that M. von Laue showed that inorganic crystals diffracted x-rays and that this effect could be used to measure interatomic distances. The experimental setup used by von Laue and his collaborators was to shine a narrow beam of x-rays through a crystal of copper sulfate toward a photographic plate. After development, the plate showed a pattern of dark spots showing where the x-rays had hit compared to the background where no x-rays hit. A pattern is exactly what is expected when waves are diffracted by a diffraction line grating or by a periodic, three-dimensional array of solid bodies such as atoms. The major requirement for diffraction is that the wavelength of the radiation must be about the same as the spacing between the solid bodies, in this case, atoms. Diffraction effects were a well-known characteristic of waves including water, sound, and light waves. Therefore, von Laue's experiment was a proof that x-rays, which were not well understood at the time, were actually electromagnetic waves like visible light and that also interatomic spacing could be measured by x-ray diffraction.

Almost immediately after reading Laue's paper, L. Bragg formulated his famous law and by 1913 solved the crystal structures of the simple cubic compounds NaCl, KCl, KBr, and KI.

Bragg visualized a crystal as made up of repeating stacked sheets of atoms, with comparable atoms equidistant from each other in and between the sheets. This analysis led to his law that the wavelength of the incident x-ray beam, L, is proportional to the distance, d, between atomic sheets, times the sine of the angle, theta, between the incident x-rays and the plane of the sheets, or algebraically, nL = 2d sin theta. In this formulation, n, is a small whole number relating to diffraction from differing sets of planes. Laue's and Bragg's work led to an explosion in analyzing the arrangement of atoms in crystals.

Again during the early 1900s, a long series of critical steps toward an understanding of DNA were made by a number of outstanding scientists. A short list of these contributors would include: Beadle, Ephrussi, Tatum, Avery, Griffith, Delbruck, Luria, Hershey, Astbury, Pauling and Chargaff (Edey and Johnson, 1989). But they were like boxers fighting an invisible foe. A bare outline of history never tells of the fog a scientist fights through at the frontiers of science. But the results always seem obvious after they are published. However, the data began to merge slowly into an idea of what DNA really represented.

Unfortunately, x-ray diffraction patterns from organic compounds are much more complex and difficult to interpret than simple inorganic compounds like NaCl. In organic materials the x-ray diffracting unit is not the atom alone but also groups of atoms and these give relatively diffuse x-ray point patterns. We should note also that a diffracting atom is not a dimensionless point in space but somewhat smeared out due to its natural vibration around a fixed position. This causes the diffraction spots to be fuzzy. Progress was slow, but by the late 1940s some improvements had been made in time for Watson and Crick to get a hint that the basic structure of DNA was a helix.

The rough consensus on DNA in 1950 was about as follows:

1. Pure crystalline DNA was not a protein.

2. DNA was made of three subgroups: a phosphate, a sugar, and a base. Its chemical composition was known.

3. DNA manufactured amino acids and proteins.

4. DNA can replicate itself.

5. DNA can replicate its host, i.e. individuals like us.

6. For every Adenine (A) there was a Thymine (T); for every Guanine (G) there was a Cytosine(C). These are all nucleic acids — called bases in the DNA vocabulary.

7. The molecular arrangement of DNA probably had the structure of a double helix.

The Watson and Crick Model of DNA, 1953

This was the situation when James D. Watson arrived at the Cavendish Laboratory at Cambridge University in the fall of 1951 to work on proteins. He had studied biology and genetics at the University of Chicago and was doing postdoctoral work in biochemistry in Copenhagen. But he felt no great attraction toward any particular scientific field until he attended a lecture by Maurice Wilkins on the use of x-ray diffraction techniques to determine the structures of organic materials. Toward the end of his lecture, Wilkins projected an x-ray image of DNA on the screen. It was a diffuse and blurred picture but it set off a chain of thought. Could x-rays determine the structure of DNA (Watson, 1968).

Watson obtained a transfer to the Cavendish Laboratory headed by Sir Lawrence Bragg. The laboratory was world renown as a center for crystallographic research. Almost immediately Watson ran into a torrent of ideas in the form of Francis Crick, a 35-year-old physicist who had switched to biology and was still procrastinating about finishing his Ph.D. These two made a team. They were both committed to the structural analysis of DNA. Watson knew about genes; Crick some physics, some chemistry, some crystallography, and a dash of everything else.

A problem for Watson and Crick was that Wilkins' research group at King's College in London was the only laboratory funded in Britain to work on DNA. But Wilkins was not overly committed to DNA; he had assigned the work to an associate, Rosalind Franklin, who now considered DNA to be her exclusive domain. The result was that Watson had to get the latest information on x-ray DNA data by roundabout means.

Finally an arrangement was made that would allow Watson and Crick to build mechanical models of possible structures for DNA. This is the popular tinker-toy method of analysis in which atoms are represented by little balls and chemical bonds by rods holding the balls together. Pauling used the method as early as 1937 to demonstrate the structure of proteins and even considered a helical structure at one point.

With the early information Watson obtained about Franklin's x-ray diffraction work, he and Crick built their first model, a three-strand helix with the bases on the outside. When Franklin and Wilkins were shown the model, they pointed out serious defects that could not be overcome.

Later, while examining Franklin's most recent and best x-ray pattern of DNA, Watson realized what Franklin had long known: DNA was unquestionably a helix. The facts about DNA were falling into place. Watson and Crick began their second model. This time they put a double helix backbone made up of alternating phosphate and sugar groups on the outside with the paired bases A-T and G-C (A-T stands for adenine-thymine and G-C for guanine-cytosine) hanging oppositely in between the two helixes. Each base is chemically bonded to one backbone strand on one of its ends, but what holds the base pairs together where they meet in the interior of the double helix? The answer was the so-called hydrogen bond.

Most chemical bonds are either covalent — they share electrons, or ionic — they exchange electrons. The hydrogen bond does neither; it forms when two different atoms, such as hydrogen and oxygen or nitrogen, are held so close together by the surrounding molecular structure that their electrons repel each other to such an extent they deform the electronic clouds around both atoms. Atoms of this type are electrically polar (i.e. positive or negative) and their opposite ends attract. The hydrogen bond forms between bases when the negative end of one base attracts the positive end of the other base. Such bonds are relatively weak, they spontaneously rupture at about 90° C (194° F) but they hold the paired helical DNA strands together. They also allow the two strands to be easily separated or unzipped during replication without breaking the stronger chemically bonded helices of the DNA molecular backbone.

The new model was a complete success. Crick and Watson published their first paper on the structure of DNA, April 19, 1953. Together with Wilkins, they received the Nobel Prize in 1962. Rosalind Franklin died of cancer in 1958, aged 37. Her cancer was undoubtedly due to excessive x-ray exposure during her x-ray diffraction work.

On Deoxyribonucleic Acid, DNA

DNA is the basic chemical molecule that controls how every individual of every species is conceived, develops to maturity, and maintains itself during its lifetime. The complexity of what it does staggers the mind. It is truly the basis of life, as we know it. What follows is a very short account of how DNA is constructed, how it reproduces itself, how it transcribes itself into a carrier of information, and finally how the information is translated to build the proteins in our bodies from only twenty different amino acids.

The three dimensional structure of DNA is a helix, similar to a coil spring. Actually it is a double helix. If you hold together two strands of soft wire, wrap them evenly around and down a pencil to one end, then remove the pencil, you will have a good model of a DNA double helix. Figure 1 shows it in two dimensions.

For even a modest understanding of DNA, we must first find our way through a thicket of symbols and figures. DNA is constructed of only five atomic elements: C — carbon, O — oxygen, N — nitrogen, H — hydrogen, and P — phosphorous. These are arranged in molecular groupings: 1, a phosphate, 2, a sugar — deoxyribose, and 3 one of four nucleic acid bases: A — adenine, T — thymine, G — guanine, and C — cytosine. Uracil-U is sometimes substituted for thymine. The combination of these three: 1 phosphate, 2 sugar and 3, a base (adenine in Fig. 3) are called nucleotides. Planar views of the seven basic molecular groups are shown in Figure 2. In these figures, the points of the hexagonal and pentagonal rings of the nucleic acids are carbon atoms except where a nitrogen atom, N, is inserted. The connecting lines are chemical bonds, four for carbon and three for nitrogen.

The continuous strands that make up the backbone of the double helix are formed of alternating phosphate — sugar-phosphate-sugar — groups. Figure 4 represents a modular block view across a short segment of a DNA double helix. Note that A and T are bonded with two hydrogen bonds, whereas C and G are bonded with three.

DNA Replication

DNA replicates by progressively breaking the hydrogen bonds between the base pairs A-T and G-C with the help of an enzyme called DNA helicase. Enzymes are small proteins that carry out or facilitate chemical processes inside the cell. As the double helix is split apart, nucleotides (Fig.3) are brought in to build two complementary strands (III and IV in Fig. 5) against the template of the original two strands (I and II in Fig. 5). Wherever there is an A on the original, a T with its constituent phosphate and sugar groups forms a hydrogen bond with it and similarly with G and C. This is the complementary principle that describes base pairing. The original double helix strands (I and II) are now replicated in the form of two new DNA helices (III and IV). The replicated helices finally combine into a double helix and function independently exactly as the original double helix of DNA.

DNA Transcription

Transcription is the process whereby DNA creates a messenger that can carry information on how to build proteins. The carrier is mRNA or messenger ribonucleic acid. RNA is similar to DNA except the sugar ribose is used in place of deoxyribose and the base uracil — U takes the place of the base thymine — T (fig. 2). In addition RNA is normally a single helix and relatively short as compared to DNA. RNA is thought to have originated in the earliest forms of replicating molecules.

The process begins when the enzyme RNA polymerase attaches to and splits open a double helical strand of DNA. Nucleotides are arranged by other enzymes to form a complementary RNA copy of the bases of one strand of the open DNA (the template strand). The sequence of bases on the RNA is therefore similar to the unused strand of DNA. The base thymine-T is replaced by the base Uracil-U in the new mRNA helix.

DNA Translation

If messenger RNA is going to carry information to build proteins, what is the information made of and what will it do? Four single nucleic acid bases alone (A, T (or U), C, and G) cannot code for or designate the twenty different amino acids that make up the protein structure of our bodies. Each base would have to stand for five different amino acids. But if we take the different possible sequences of two bases, we can get a code of sixteen: symbolized by TA, GC, AG, AC, TG, TC, AA, TT, GG, CC, AT, CG, GA, CA, GT, and CT. That is still not enough to specify uniquely each of the twenty amino acids. However, if we take three bases at a time and consider the possibilities, we can have 64 different code symbols — more than enough. This triplicate code is made up of "codons" like AUG, UUA, CGU, and AAA. Fig. 7 gives the complete set of 64 codons together with their designated amino acids such as glycine or alanine (U is a duplicate for and replaces T in this Figure). There are few exceptions to these relationships. Note that some of the three letter codons, designate the same amino acid and UAA, UAG, and UGA are used to terminate strings of codons much like a period at the end of a sentence.

Recall that all the proteins of our body are made of only 20 of the thousands of different amino acids that exist. How can these few create the incredible variety of proteins that make up living tissue? The brief answer is that proteins are made of chains of the 20 amino acids in different combinations and sequences and can fold up into unique convoluted structures that are the most complex and sophisticated molecules that exist both structurally and functionally. The folding is brought about by a combination of hydrogen bonds, ionic bonds, and van der Waals forces between the atoms of amino acids and their neighbors.

Amino acids (Fig.6) have a distinct structure quite different from the nucleic acids of DNA. They have 4 typical groups attached to a central carbon, called the alpha carbon. The groups are a hydrogen atom, a basic amino group, an acidic carbonyl group, and an R group which can be any one of a great variety of chains and functional groups. Acids and bases strongly react with each other, often with the release of considerable energy. Because one end of the molecule is basic (the amino group) and the other acidic (the carboxyl group), an individual amino acid can react with any other amino acid to form long chains. This acid-base reaction splits out a molecule of water forming a new molecule that also has base and acid groups at opposite ends. The process can continue indefinitely as long as more amino acids are available. The new bond between amino acids is called the peptide bond and is a non-rotating bond in a protein chain. The overall reaction is much like those that produce synthetic polymers such as nylon, polystyrene, and polyethylene from a series of small monomers.

Here we can pick up the messenger RNA we left a few paragraphs ago. Each three-letter codon on mRNA specifies a particular amino acid. For instance, a sequence like GGGAAGGCC is broken down into the codons GGG, AAG, and GCC. From Figure 7, these codons stand for the amino acids glycine, lysine, and alanine. The mRNA gathers the amino acids in the sequence shown to build the first section of a protein. Long strands of messenger RNA break apart to make many sequences simultaneously at various sites. If needed, other mRNA molecules join in to produce more segments of the protein and put them all in order. Positioning and ordering is brought about with the help of many more different enzymes. Finally, to overwhelm us completely with the complexity of the process involved, proteins are extremely difficult to unfold and analyze, resembling nothing so much as a tangle of ribbons or a bag of worms.

In the last few pages we have come through a thicket of complexity, change, processes and exceptions that strain credulity. But what we have been through is a very simplified and abbreviated version of the known facts (Alberts, 1998).

The recent understanding of the structure and protein forming abilities of DNA are an overwhelming and thorough confirmation of the theories of evolution and natural selection. The processes involved in DNA replication indicate in molecular detail how mutations occur and individual variation comes about.

Lord Kelvin and the Problem with Time

Earlier we considered the problems scientists were having in determining the age of the earth. There were two opposing camps; one favored Uniformitarianism (slow changes on an old earth) the other Catastrophism (rapid changes on a young earth). Lyle and Darwin were convinced they needed long periods of time for the vast changes they observed. Into this argument a world famous physicist and mathematician, William Thomson (1824–1907) later known as Lord Kelvin, took a disruptive step. In 1854 he delivered a paper "On the Mechanical Energies of the Solar System." that proposed the heat energy of the sun and earth were originally due to the gravitational collapse of gas, dust, and meteor swarms into very hot bodies. Under the force of gravitational attraction dust and meteors collide and coalesce, converting energy of motion into heat energy in accordance with the law of the conservation of energy or the first law of thermodynamics.

Similar ideas had been around for some time as part of the nebular hypothesis of Immanuel Kant and Pierre Laplace as well as Hermann von Helmholtz. Lord Kelvin added mathematical calculations, however, detailing the rate of earth's heat lose into space and heat gain from the sun. This allowed him to estimate the age of the earth. With each paper he wrote over the next fifty years the age varied, but overall he claimed the sun had illuminated the earth for less than 100,000,000 years and certainly not more that 500,000,000 years. Even more importantly, the earth had not cooled enough from a state too hot to allow life to exist, that is, about +/−200 degrees centigrade, for more than the last 20,000,000 years.

Since Kelvin supported these claims with all the powers of his eminent position, he attracted many followers, particularly among physicists. Geologists and biologists, although unable to refute Kelvin mathematically, maintained that their observations of natural processes could never be reconciled with such short periods of time. Kelvin disregarded their data and their views. His attitude was that geology ignored the established laws of physics and therefore geological data was of no importance. In 1868 he delivered a polemic against long geologic time and by inference against the theory of evolution (Burchfield, 1975).

Scientists were forced to waste many decades trying to get out from under Kelvin's 20–100 million year limits. In particular, the geologists tried to calculate the age of the earth by quantifying various geological processes. The rates of erosion, sedimentation, and salt concentration in seawater were the subject of many scientific papers; but the variability of these natural processes was too great for conclusive results.

Kelvin had made important contributions to many fields of science including thermodynamics and electromagnetism, but he was violently opposed to Darwin's evolutionary theory. His objections to the natural selection mechanism were three: first, natural selection did not account for the origin of life; second, it required more time than the basic laws of physics would allow; and third, it replaced the obvious evidence of design in nature with unacceptable chance happenings. Since the first and third objections could not be studied definitively by the scientific methods of the day Kelvin concentrated his attack on time.

Kelvin made at least two fatal errors. First, he thought since then current laws and theories of physics were on his side, the other side was ipso facto wrong and the data of geology and biology were bogus and could be ignored. Second, he did not adequately consider that new phenomena might be discovered that would supply the basis for a much more ancient universe. In fact, when radioactivity was discovered, he refused to accept it as an important heat source. As late as 1906 he wrote a series of letters declaring that universal gravitation remained the only possible source of the vast energy radiating from the sun that heated the earth. From the historical evidence Kelvin never changed his mind.

Meanwhile, in early 1903, Pierre Curie and Albert Laborde found that radium salts constantly give off heat. About the same time Ernest Rutherford, Frederick Soddy, and Howard T. Barnes proved that radioactive materials radiate heat proportional to the number of alpha particles they emit. In 1904, Rutherford gave a lecture in London (which Kelvin attended) proposing radioactive decay of the elements as a continuing source of thermal energy and pointed out that this could account for the moderate temperatures over the long time periods needed for evolution to take place. Within a few years most physicists and geologists were convinced that radioactivity was a major source of terrestrial heating.

Not only were the radioactive elements considered a source of heat, but evidence was accumulating that the radioactive decay process could be used directly to determine the age of the earth. Over the next few decades many of the details of radiometric dating were established. In 1921 and again in 1922, international symposia on the subject, accepted 1.3 billion years as the age of the earth. The US National Research Council issued a report in 1931 citing the radioactive decay of uranium into lead as the most accurate dating method and gave 1.5 to 3 billion years for the age of the earth. This signaled the effective end of the argument about the main source of earth temperatures, while the estimated age of the earth has steadily increased as older rock strata were discovered and dated. The age of the Earth at present is estimated to be 4.6 billion years.

Tautology and Falsifiability

The theory of natural selection has been criticized as a tautology and therefore without content (Johnson, 1991). Tautologies are statements with no exceptions and are necessarily true. A typical tautological statement such as: all bachelors are unmarried, tells us nothing more about bachelors than what is already contained within the definition of the word bachelor. Similarly, the critics claim the phrase: survival of the fittest means only that those who survive are able to survive. This is a tautology and therefore does not extend our knowledge about the causes of survival. But neither does the label tautology prove the statement is false nor that the process of natural selection is false. The latter is what the critics really want to prove by their charge of tautology. This cannot be done since a tautology cannot prove truth or falsity; it can only be charged with redundancy.

According to the philosopher, Wittgenstein, logical truths are all tautologies (O'Connor and Carr, 1982). Tautologies can be definitions, syllogisms (the premises logically necessitate the conclusions), laws of logic, logical inferences or equalities. They have no additional truth value. The problem is that they are redundant; they tell us nothing new. If the survival of the fittest is a tautology, it proves nothing about the truth or falsity of the statement or the process of natural selection.

We should note that, the survival of the fittest is not Darwin's phrase. It is a phrase that was originated by Herbert Spencer, one of Darwin's supporters, and was catchy enough to capture the popular imagination (Dennett, 1995). In addition, the phrase is not always true as required by the definition of tautology. In a moderate environment almost all individuals survive and no individuals survive in an especially harsh environment. Being the fittest member of a species may or may not help survival, depending on the changing environment over which fitness has no control. There are always at least two completely independent factors operating in natural selection — individual and environmental variation. No individual can know or prepare itself for the environment it will face (Michod, 1995) although it should be noted that birds molt in the summer so as to be ready for winter whereas humans put on fur coats or start a fire because their DNA does not furnish them with more hair.

Darwin's definition of natural selection was quoted earlier from the introduction to *The Origin of Species*. In this quotation, both the individual and the environment are considered as variables. If we also include reproduction, inheritance, time, and chance as part of natural selection, we end up with at least six variables in the process. Can anyone seriously claim that a two or six variable process is the same as or equivalent to a statement such as "all bald heads are hairless"?

The word "fittest" is a superlative and means: the best one out of a group of three or more, although it is often extended to mean groups. A better word would be "fitter," since this could specify a class of individuals. A still better phrase is "the survival of those fit enough," which is also a tautology but truer to the facts. All individuals living at this moment are obviously fit enough to have survived to this moment.

As noted, in a benign environment all individuals might survive; in an extremely harsh environment no individuals survive. The Grants (1989), in their studies of the cactus finches on a waterless Galapagos island, found there were environments in which few could survive and others in which almost all survived depending on the annual rainfall which in turn controlled the kinds of plant seeds available.

Many other scientific concepts can be cited for the sin of a tautological nature. For instance, particles with mass are held together by the mutual attraction of gravity; or gravity is the apparent attraction of masses in a space-time curvature; or the mutual attraction of masses is gravity, etc. Again it must be accented that a word like gravity is defined by observation not by deductive logic. Gravity is defined by observations of how two masses move in relation to each other. To repeat, tautologies are not statements about truth or falsity since, by definition, they are always true because of the grammatical structure of language.

The issue of "falsifiability" has also been raised as showing the unscientific nature of evolutionary theories (Johnson, 1991). Karl Popper was the first to accent falsifiability as the major determinant of the truth or falsity of a scientific theory. His thesis was to test a theory by looking for instances where the theory does not hold. This is exactly what scientists have always done. They spend countless hours searching for information that contradicts or confirms an emerging hypothesis. They need no prodding from Popper. His point was that scientific theories are arrived at by the inductive method and that no amount of confirming information can prove its truth since endless time and events are available for disproof. How many confirming experiments must be made before Popper is satisfied? To say scientists should use most of their time hunting for exceptions to their theories is unreasonable, if not ludicrous. Why not devote time looking for more data and theories? There are always independent colleagues who willingly seek additional confirming or contradicting evidence. However, Popper has also written the following, of which only three words need be deleted, namely "scientific" and "one of" (Popper 1944):

> Scientific knowledge and human rationality that produce it are, I believe, always fallible, or subject to error. But they are, I believe, also the pride of mankind.... Scientific knowledge is, despite its fallibility, one of the greatest achievements of human rationality.

In response to Popper's criticism and praise we can only say welcome to the real world of probabilities. Most of us would be happy with a theory that passed a thousand different tests and failed only one. We would call the one, a paradox, a misrepresentation, an anomaly or an error, and pass on to our next triumph. This is exactly how scientists live, on the edge of events that can weaken or destroy a month, a year, or a lifetime of endeavor.

Popper's thesis evidently requires that only deductive logic be used to prove a truth. Unfortunately, as we saw in the chapter, "The Search for Truth," deductive logic, the syllogism for example, only proves what we already know from the original premises. There is no possibility in deductive logic for a creative leap into new conceptual regions, which is the most important characteristic of scientific observation and experiment. We can be sure that a new theory for a major problem in physics today (a theory to rationalize the differences between relativity and quantum mechanics theory, for instance) will not be found by deductive logic alone.

A scientific experimenter uses his knowledge, skill and logical abilities in an effort to achieve greater understanding. When scientists conceive new theories based on their experiments, the first thing they do is to test the theory in all the relevant situations which can be imagined, including both reinforcing and destructive tests. Changes and adjustments are often necessary. Alternate theories are carefully analyzed. Even so, nine out of ten theories fail during these formative stages. Self-criticism is a required trait for a productive scientific life. Also important is the knowledge and skill of colleagues when results are published, or described at a technical meeting, or claimed in a patent. A new theory is always presented before the scientific community for acclaim or disdain. There can be no more anxiety-laden moment in a scientific life than when a new theory is presented to a peer group. To summarize, falsifiability criteria are applied long before a scientific theory ever becomes common knowledge, and publication opens the theory to criticism by the most competent specialists in the field as well as anyone else who wants to join in.

In the case of evolutionary theory there are many major tests conforming to Popper's falsifiability thesis that have been successfully met since Darwin's death in 1882. One is the disproof of Lord Kelvin's restrictions on time; a second is the proof of a mechanism (DNA) for the inheritance of variations which have passed the filter of natural selection; and a third is the illumination of the molecular mechanisms which produce variations among individuals in the first place.

Lord Kelvin's time limitation of 20–100 million years for the age of the earth was wrong and confused the issue for at least 50 years. The discovery of radioactivity in 1896 led to the recognition that the earth was roughly 4.6 billion years old and that the earth's temperature was maintained by both radioactive decay within the earth and radiation from the sun.

The elucidation of the mechanisms which control heredity was reduced to basic atomic and molecular processes with the discovery of DNA, the double helix, RNA, etc. The new discipline of molecular biology has laid bare the complexities of living structures with a depth and clarity unthinkable fifty years ago. The 21st century will certainly see applications of molecular genetic information that will transform life as we know it.

The variations that characterize life have also been shown to have a molecular basis. The processes of mutation, meiosis, crossing over and fertilization all have a part in originating changes among individuals.

These three examples show clearly that Popper's falsifiability criterion have been met by the theory of evolution. Whether that proves anything or not, is another matter. The necessary scientific phenomena and techniques had not yet been discovered to solve the problems of time, variation, and inheritance when Darwin published his theory in 1859. Darwin was well aware of the problems and made several attempts to overcome them within the confines of the science of his day. Would it have been wise for Darwin to throw his theory into the trash bin because techniques were not yet discovered that would confirm his theory? His data and analysis required long periods of time, explanations for variation, and inheritance of changes in traits. Theories are models against which we can test data and concepts. They are extremely important for science whether they last forever or not. They are important if they are only truer than other data or concepts at the time.

Here is the fatal flaw in Popper's theory: he is mesmerized by absolutes. If there is one scientific datum that falsifies a theory but much data that supports it, can that disprove the scientific basis of the theory as evolution's critics maintain? Falsifiability proves or disproves nothing except perhaps that the deciding information is not yet available. A single bit of falsifying information often has no more importance than a single bit of supporting evidence. Experimental and analytic errors have been known to occur! We must keep our options open and our judgment clear of suppositions. Scientists use a theory as far as they can take it, often throwing light on subsidiary problems along the way. The final arbiter is always time and the critique of the scientific community. We can be sure that a faulty theory will be disproved or replaced as the store of scientific knowledge increases.

Darwin's Problems

Why did Darwin wait so long to publish the results of his investigations into evolutionary processes? After all, the broad outlines of his theory were in place after he read Malthus' Essay on the Principle of Population September 18, 1838 (Mayr, 1991). In May 1842 he set down his ideas in a 30-page outline and in 1844 expanded the outline to 189 pages which his friend Hooker read. We cannot be sure of the answer to our question, but we can be sure that among the major reasons was Darwin's dissatisfaction with some parts of his theory. He, more than anyone, realized there were areas that were controversial, muddled, or missing.

In particular, there were two problems that would bedevil Darwin until his death in 1882. First, there was still the problem of time. In 1862 Lord Kelvin, the distinguished physicist who devised the Kelvin scale of temperature, determined the age of the earth as 20 to 100 million years from estimates of cooling rates. This was much too short a period for evolving life to have come to its present level of complexity. Second, there was the problem of inheritance. How do individuals inherit traits from their parents? Darwin had tried "blending" and tiny, invisible "gemmules" as mechanisms for inheritance, but neither idea passed the filter of the facts. Perhaps the weakness of his theory in these two areas were the major reason he did not publish until forced to by the sudden appearance of Wallace's paper in June, 1858 that also described common descent and natural selection.

Bibliography

Alberts, Bray, Johnson, Lewis, Raff, Roberts, and Walter, 1998, *Essential Cell Biology*

Burchfield, J. 1975, *Lord Kelvin and the Age of the Earth*

Dennett, D. 1995, *Darwin's Dangerous Idea*

Edey, M. and Johnson D. 1989, *Blueprints: Solving the Mystery of Evolution*

Grant, B. and G. 1989, *Evolutionary Dynamics of a Natural Population*

Gribbin, J. 1985, *In Search of the Double Helix*

Margulis, L. and Sagan, D. 1995, *What is Life?*

Mayr, E. 2001, *What Evolution Is*

Michod, R. 1995, *Eros and Evolution*

O'Connor, D. and Carr, B. 1982, *Introduction to the Theory of Knowledge*

Popper, K. R. 1968, *Conjectures and Refutations*

Watson, J. 1968, *The Double Helix*

Chapter 12. The Micro Evolution of Life

A need for meaning and purpose in life is characteristic of self-aware, conscious beings. It is often expressed as an aim, goal, or felt as a need for understanding, significance and love from others. All living creatures have the purpose to maintain life and reproduce. Our need for meaning is directly related to these purposes.

The meaning of life and the meaning of evolution are similar in that they are parallel and interrelated. Meaning arises with the beginnings of life on Earth 3.5–3.8 billion years ago. Evolving life embraces continued life and procreation as vital goals and purpose. Life has purpose and purpose gives direction and meaning.

Even the most primitive creatures need such things as food, air, and water. They interchange mass and energy with their environment and can exist only in environments that fill these needs. With life, purposeful action to fill needs first appears in the universe. It is we living individuals, consciously or not, who decide what purpose is, what has it and what does not. As always, it is the individual creature that makes the choice and reaps the gain or pain.

The Emergence of Purpose

The inanimate universe has no purpose; it is simply there. It consists of material particles and energy obeying their fixed chemical and physical natures while following a path of purposeless change. Particles and photons did not set out with the purpose or goal to create life; they simply followed the endless, interactive, and determined changes characteristic of the inorganic world. After the "Big Bang" occurred about 13.7 billion years ago and the universe expanded for another 10 billion years, replicating entities emerged from chance encounters of atoms obeying their fixed chemical natures. These would continually change and evolve erratically for three and a half billion more years before we appeared 0.2 to 1.8 million years ago.

THE DOUBLE HELIX

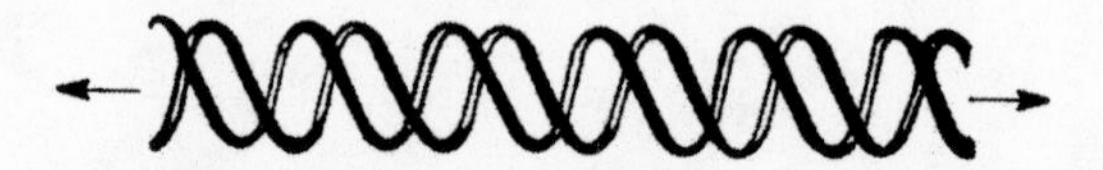

FIG 1 A DOUBLE HELIX

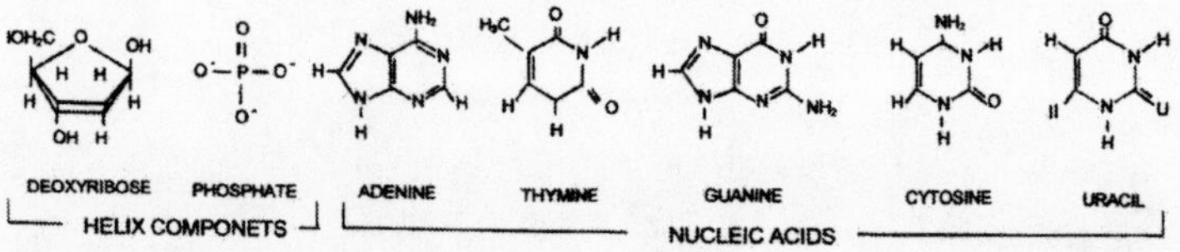

DEOXYRIBOSE PHOSPHATE ADENINE THYMINE GUANINE CYTOSINE URACIL

HELIX COMPONETS NUCLEIC ACIDS

FIG 2

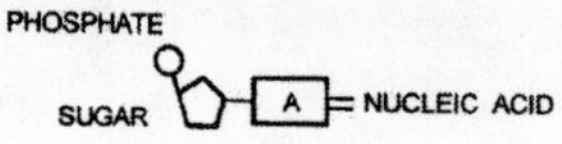

PHOSPHATE
SUGAR A NUCLEIC ACID

A NUCLEOTIDE
FIG 3

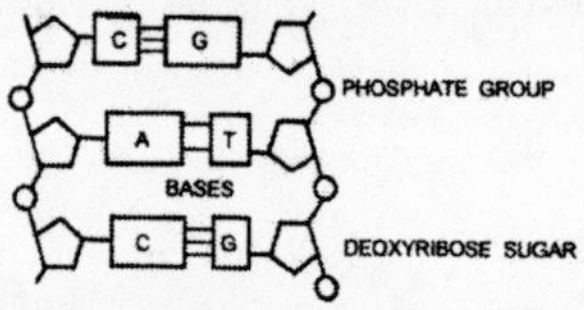

C G
A T
C G
PHOSPHATE GROUP
BASES
DEOXYRIBOSE SUGAR

MODULAR VIEW
FIG 4

Living creatures were designed (metaphorically) by their predecessors and the creatures themselves over billions of years. They are the ones who endured the events described by the natural selection process, the onslaught of harsh environments and selective deaths that eventually led to newly capable creatures. Living forms, in essence, are their own continuing creations. They are the individuals who struggle to live and reproduce. If successful, the individual passes life onto others before dying; if not, the individual has participated in the joys and hazards of its species. In either case, they have continued a life that varies, endures briefly, and dies. To deny individuals the truth of this aspect of self creation denigrates all life, living or dead.

THE DOUBLE HELIX

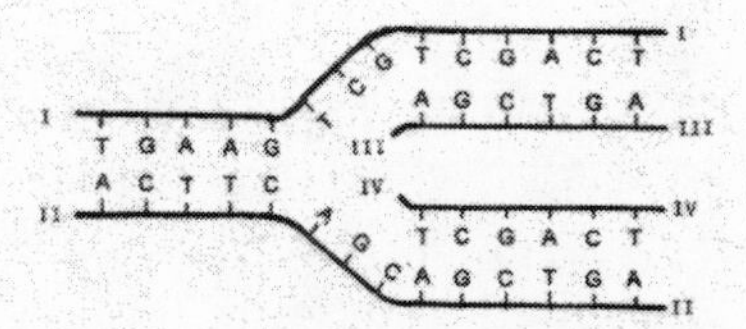

FIG 5 REPLICATION

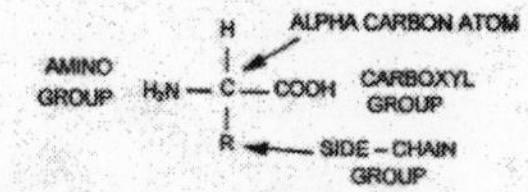

FIG 6 AMINO ACID

	U	C	A	G
U	UUU, UUC — PHENENYL-ALANINE UUA, UUG — LEUCINE	UCU, UCC, UCA, UCG — SERINE	UAU, UAC — TYROSINE UAA, UAG — TERMINA-TION	UGU, UGC — CYSTEINE UGA TERMINATION UGG TRYPTOPHAN
C	CUU, CUC, CUA, CUG — LEUCINE	CCU, CCC, CCA, CCG — PROLINE	CAU, CAC — HISTIDINE CAA, CAG — GLUTAMINE	CGU, CGC, CGA, CGG — ARGININE
A	AUU, AUC, AUA — ISOLOUCINE AUG METHIONINE	ACU, ACC, ACA, ACG — THREONINE	AAU, AAC — ASPARAGINE AAA, AAG — LYSINE	AGU, AGC — SERINE AGA, AGG — ARGININE
G	GUU, GUC, GUA, GUG — VALINE	GCU, GCC, GCA, GCG — ALANINE	GAU, GAC — ASPARTIC ACID GAA, GAG — GLUTAMIC ACID	GGU, GGC, GGA, GGG — GLYCINE

FIG 7 TRIPLICATE CODON CODE

The Facts of Evolution

Since Darwin's time, major new facts about the evolution of life and its relationship to the physical world have been uncovered. These include the atomic structure of life, the measurement of time by radioactive decay rates and the red shift of light, the inherent instability of matter, the inheritable instructions for life in molecular DNA, and the many near-extinction events that accompanied evolution on Earth.

The new facts confirm and amplify the original Darwinian theory of evolution. Confirmation by these new, often unrelated, scientific facts and theories show the flexibility and adaptability of evolutionary theory as well as its successful passage beyond the "falsifiability" criteria favored by Karl Popper and others (Johnson, 1991).

We often embrace ignorance if only to preserve our myths. But the time for fanaticism, denial, ignorance and indifference is over. The facts of evolution should be known by all. We should act on the facts of evolution now; to do otherwise is to flirt with extinction.

Some say that evolution is only a theory, it is not a fact. Of course evolution is not a fact; it is millions upon billions of observed facts that coherently explain a set of related phenomena by scientific test methods.

Timely Design, Result, or Outcome

The central factual observation of the universe is its continually changing nature. The events that characterize evolution and the natural selection process also exhibit continual change. The type of design that is characteristic of evolution is timely design. It is a continuing activity that operates on every creature that ever existed over every moment of its lifetime since the beginnings of life on Earth. Evolution is continuous design and redesign, ad infinitum. Design, as normally used, implies that a thing or machine is finally completed and will henceforth carry out a fixed function. But the universe is undergoing continuous change; so which universe or which transient life form had a fixed design?

As noted earlier, natural selection "selects" nothing. Similarly, timely design "designs" nothing. The words "select" and "design" are heavily burdened with inherent purposeful meaning that misleads the unwary. Perhaps we should use the phrase "timely result" or "timely outcome" to define the idea of time-based events. The time-based application of natural selection results in living creatures that have attributes enabling them to live in a particular environment at a particular time. Such events occurred with every creature at every moment of time since life began some 3.8 billion years ago and are occurring now.

The Measurement of Time

Life evolves in the universe. We had many predecessors; the earth is littered with our fossils. Life evolved on Earth over a period of at least 3.5 billion years according to the fossil record.

Our ability to measure the dates of past events rather than sequences alone became possible relatively recently, during 1900–1930 with the discovery and confirmation of atomic radioactive decay rates and the red-shift of light (see "Scientific Reality"). Atoms that decay radioactively into other atomic species do so in exact sequences and time periods. The scientific law is that one half of an atomic species decays into different atomic species in a definite interval of time (the half-life). For instance, one half the individual uranium atoms in a mineral sample each decay into eight helium atoms and one lead-206 atom over an average period of 4.51 billion years. By analyzing uranium minerals for the number of their uranium and lead-206 isotopes, we can calculate the age of the mineral and the age of the rock strata where it was found. Although we cannot determine when any particular uranium atom decays, we can determine statistically how many atoms decayed in a small sample (the number of atoms in a tenth of a gram of uranium, for instance, would be about 6 followed by 21 zeros). Since fossils are often imbedded in rock strata, if the strata can be dated by radioactive decay rates, the age of the encased fossils can be determined. To take advantage of this recent information about dating, all previous human knowledge and history should be brought into conformity with the more accurate scientific laws of time measurement, dates, and events. To do so will bring about a vast reinterpretation of past events. In fact, the geologist's designations such as Triassic, Phanerozoic, and Hadean, etc., although historically useful, are confusing, and should be replaced with the numerical dates and intervals that all can understand without specialized training.

Life is made of atoms. We come from starbursts and sunlight — matter in the process of change. The material earth is made of ninety two naturally occurring different elements; another 17 or so have been created in the laboratory for brief periods. In addition there are a number of smaller particles (electrons, protons, neutrinos, quarks etc.) and photons (thermal waves, light, gamma and x-rays, etc.) that complete the known roster of mass and energy. The atoms, particles and photons found on earth are identical to those found elsewhere in the vast inorganic universe we can observe by telescopic means. We know this from the unique light spectra we receive from stars; they match those observed from the same atoms on earth. Although life is made of atoms, it is the accumulation and chemical interaction of discrete atoms into complex molecular structures, which makes possible the transformation of inorganic matter into self-replicating life. Since the possible interactive changes and permutations of 92 elements are almost endless, a selection method that preserved stable combinations would have dramatic consequences. The process of trial-and-error in the physical universe is the only selection method we know of for stable combinations. In the biological universe this process is called natural selection.

Change is the basic characteristic of the universe. Endless change is the essence of existence. At this moment, every atom in the universe is moving with respect to every other atom and every galaxy is moving with respect to every other galaxy. In addition, the universe is expanding in every direction. The processes of evolution and life also participate in this law of universal change. Each and all living cells at this moment are undergoing changes in their internal chemical composition and structures.

All life forms and their cells are related through molecular DNA and use the double helix molecular structure of DNA (deoxyribonucleic acid) to encode instructions for procreation, growth, and maintenance activities. DNA is predominately made of the five elements carbon C, oxygen O, hydrogen H, nitrogen N, and phosphorus P. Excepting oxygen and hydrogen, each of these are relatively rare. However, life forms have recently been discovered around energy and mineral rich volcanic vents on the ocean floor. These raise the possibility of other species here or elsewhere in the universe that use different energy sources, atomic elements, and encoded instructions for life and procreation.

Living creatures have evolved from earlier creatures. One of Darwin's early tenets was that living creatures are related to earlier forms as shown by the fossil record. From the DNA evidence now available since Darwin's death in 1882, it is obvious that earth creatures are related by evolution to a single source, wherever that source may be — Earth, Mars, other solar systems etc.

Natural selection results in evolutionary change. Natural selection is the evolutionary process that describes from among various individuals which can live in a particular environment and which cannot. It relies on inanimate physical events that have neither plan nor purpose. It is simply a result of physical happenings under random conditions; it selects nothing. Faulty variations of living creatures are detected and discarded much as a quality control engineer rejects parts that do not meet specifications. "Faulty" in this context means not adapted for survival in a given environment at a given time. Natural selection is a short-term process but it can yield astonishing long-term results. Actually it is not a process at all; it does nothing; it does not amplify, pause, create, encourage, improve, design, etc. etc.; it is not active. Natural selection is simply a description of the outcome of events that occur when a living creature, which is changing itself, tries to live in a changing environment over which it has no control. If it is an extremely adverse environment the creature dies or has fewer offspring; if benign, the creature lives. Natural selection, in essence, simply describes the environmental limits within which particular individuals have been able to live at particular times. Perhaps we should call it the Natural Outcome or Result in order to remove the inference of selection.

Major Physical Events

From what we know by radiometric dating and the red shift of light, there are periods in the universe and evolution that can be designated as major events. These are somewhat arbitrarily listed below. Dates are approximate:

Major Physical Events	
The universe began in the "BIG BANG"	13.7 BYA*
Heavy atoms (C, O, N, P), were created in supernova	<11 BYA
Solar system evolved	4.5 BYA
End of meteor showers on Earth	3.8 BYA
Major Biological Events	
Replicating molecules	3.9 BYA
Single celled prokaryotes, mitosis	3.7-3.9 BYA
Photosynthetic bacteria, first oxygen	3.6 BYA
Oldest fossil evidence, bacteria	3.5 BYA
Oxygen abundant — aerobic bacteria	1.8 BYA
Eukaryotes, multi celled, meiosis, bisexuality	1.5 BYA
Cambrian era — animals, plants, fungi appear	500-570 MYA
Hominids evolve	2-4 MYA
Modern humans appear	.02–1.8 MYA
*BYA= billion years ago, M= million; T= thousand	

The first four of the above lists cover the creation of the universe and solar system; the next five are the period of single-celled life forms, the prokaryotes. With a huge, poorly defined overlap, the last four are from the period of multi-celled life, the eukaryotes. The period from 3.6 to 1.5 BYA, about 2 billion years, contains momentous changes, most importantly an increase in atmospheric oxygen, which forced a change to oxygen metabolism. The result today is a vast mixture of multi and single celled creatures trying to live in harmony with *Homo sapiens* (Schopf, 1992, 1999).

The single-celled era encompassed the creation of life, DNA, the cell, mitosis, and the beginnings of Earth's oxygen atmosphere produced by photosynthesizing bacteria. The multi-celled era saw the evolution of, bisexuality, meiosis, plants, animals, fungi, hominids, and humans. The interval in between, the conversion to oxygen metabolism, is unrelenting trial and error, turmoil, life and death.

The Accumulating and Micro Nature of Evolution

Earlier we discussed some of the steps through which eyes evolved to reach their present state of complexity. The list of steps included light sensitive molecular compounds, pinholes, lenses, and reflectors but each of these consist of unending slight changes through tens or hundreds of millions of years. When naturalists talk of evolutionary steps they are usually thinking of distinctions that are visually observable and stand out from the vast panorama of nature. But these large scale macro displays are the accumulated results of barely discernable continuous slight micro changes (Dennett, 1995). For instance, we cannot yet observe a single atom reacting with another in a living cell but we can often measure long-term concentrations of the resulting atomic reaction products.

At this ultimate micro atomic scale, what is natural selection? It involves the same two totally different events which are continually and independently changing life forms themselves. The first independent variable is the individual living creature that is fragile, varying in DNA, and can exist only under extremely limited chemical and physical conditions. Since life forms contain DNA, temperatures must be maintained below about 90 degrees centigrade to avoid damage; it must have free access to oxygen and many other elements that are soluble in water in ionic form; it requires the input of food for metabolism but shielding from high energy radiation, etc.

Living creatures exist in complex internal and external environments that are the second independent variables. Environment has none of the attributes of life and is largely chemical and physical in nature. These two, the varying creature and its varying environment at the moment, are each continually changing according to their basic natures independently of the other. If the environment changes beyond the conditions necessary for life, the living creature dies; it was not "selected."

Natural selection is "acting" now, continuously on every living cell and creature populating the Earth. Natural selection "acted" continuously on all living creatures (bacteria to humans) that ever existed on Earth over the last 3.8 billion years. Natural selection never ends, pauses, or comes to completion; it is continuous, sensitive to change, and unrelenting. The result is that living creatures tend to accumulate the slightest beneficial change. Only this endless, moment by moment application of continuous trial and error and natural selection could have produced the time based complexity and diversity we observe in life today.

Life is optimism triumphant; it is optimism corporeal. Living creatures ask for nothing but need much. Life strides on confidently, never questioning what lies ahead. It is master of the universe; certain it will find its way. In its later eukaryote phases, the DNA sometimes evolved techniques to detect danger and avoid it. But a tulip poplar tree stands tall and straight, never afraid of falling, even though the ground is littered with dead trunks, branches, and leaves; optimism embodied, indeed. We conscious creatures alone know our fate is death. The human ability to empathize, to vicariously sense the feelings and experiences of others bring us that knowledge.

Although we often ignore or deny it, we are not the purpose or objective of evolution. Evolution is not progress toward a goal, it is change for change itself. Life evolved late to consciousness and reason. The first hominids appeared only 2–4 million years ago, modern humans about 0.2–1.8 million years ago. Why did it take so long — more than 13 billion years of universal existence including about 4 billion years of living existence on Earth? Conscious and reasoning life must be a very rare event in the cosmos, considering the variety and sequence of events through which we have evolved.

Primitive types of molecular replication probably occur in the universe wherever appropriate chemical and physical conditions exist, but the possibility of other conscious creatures that we can find and communicate with, is remote. For all practical purposes we are alone simply because of the vast cosmic distances involved for observation and communication.

The Meaning of Death and Extinction

Species eventually become extinct. The fossil record gives the average lifespan of a species as one to ten million years. Considering our present dominance of the earth after only a few hundred thousand years we can foresee a long future, particularly if we can overcome the escalating increases in human population pressure and the exploitation and consumption of resources that accompanies it.

However, from the fossil record it is also clear that there have been at least five major mass extinction events. For instance, the mass extinction due to an asteroid impact toward the end of the Crustaceous Period about 65 million years ago caused the extinction of the dinosaurs as well as 75% of all marine life (Raup, 1991).

All multi-celled creatures die. The most long-lived creatures known are the bristle cone pines of the southwest United States. Core samples from living specimens have had average ages of about four thousand years. There is no evidence from the fossil record that any individual creature has lived longer. Although single celled creatures can die by accident, they normally divide into two identical individuals. In a sense, they never die; cell division (mitosis) is not the death of one life but the disappearance of one and the simultaneous appearance of two. It is a change, not death.

Apparently, nature was not able to accept or use endless continuation as a defining characteristic or necessity of life. The evolution of bisexual creatures (eukaryotes) required death. If early replicating creatures had been completely stable, there would have been no variation, no natural selection, and no evolution. Three of the essential components of DNA, carbon, phosphorus, and nitrogen, are relatively rare; C and N each make up less than 0.1% and P less than 0.12% of the earth's surface crust (Chang, 1988). A completely stable, replicating life form would soon incorporate all these rare elements from the available environment, leaving nothing for other species. Note that even with a replication rate only slightly above 2 per mating couple plus increasing longevity, we humans are in essence following an equivalent path at the present time. We are exhausting the Earth in a much broader sense. It is not scarce elements alone that are usurped but land and sea area needed by other creatures, and energy, and materials in general.

The Meaning of Life

Can we recognize human meaning and human purpose from scientific facts and their implications?

Each individual is unique. This unique nature is the cost and reward of existence and evolution. Because each individual has somewhat different DNA from all others, the processes of natural selection and evolutionary change are possible. Living individuals are by definition, complete and independent. They can exist alone or form families and flock with others in symbiotic relationships.

The individual carries the total burden of evolution. Societies and species are theoretical concepts that are human definitions for groups of individuals. Species have no life, struggle, pain or death. At extinction they simply end or disappear. We creatures of atoms live, endure, and finally die in sadness, if not anguish, alone. The individual is the key and cornerstone of life. Without individuals there would be no creation of life, no evolution, no humans, and finally no societies. One might say nature "learned" that the process of evolution could not begin or continue without simple, discrete individuals that could each adapt and exist alone.

Conflict between Individuals and Organizations

Tension exists between individuals and their organizations. From the individual point of view, the value of societies and other organizations is that they can accumulate, preserve, and transfer individual accomplishments thereby increasing the potential capabilities of all. Societies create nothing nor do they acknowledge inherent self-controlling principles. However, many individual accomplishments can only be implemented by relatively large organizations of people. Economic activity, for instance, is dependent on collective activity and the quality of orchestral music depends on the synchronous action of all its players.

However, individuals must protect themselves from the relentless expansion and diversity of organizational power. A balance must be achieved between individuals and organizations for the benefit of both. Unfortunately, human history has been largely a parade of individuals or small groups gaining control of state organizations in order to increase their personal economic, political and military power. These have always ended in collapse but at a terrible price for individuals (Diamond, J. 2005; Tainter, J.A. 1988).

Our ancestors gave us life. They struggled for and maintained life and formed a bridge of being — the gift of life to us. We will do the same. We must honor them all; if we do not, *our* lives can have little meaning.

Our yearning for excessive power belittles our ancestors. We were given no plan, no purpose, no meaning. We must find our way as our ancestors did along the optimistic but arduous trek from bacteria to early man. We alone can determine the purpose and meaning of life because we are life — conscious, knowing life; all else is fantasy.

As individuals we are alone. For the foreseeable future, perhaps forever, we are alone in the universe. We were born alone and die alone. Self, family, and friends form the core of personal life and help raise us to a full level of humanity but, in essence, we are alone. No one else can breathe your breath nor pump my blood. No one can love or be loved but individuals. We can look nowhere beyond ourselves for a plan and purpose that will give our life meaning. We must determine and follow the path to enlightenment making our imperatives extend in reverence to all life including self, family, friends, societies and our companions in nature.

The Natural Selector

We are the Natural Selector. Because of the exploitation of the discoveries of science and the dominance of technology, our collective behavior now controls the fate of all living creatures. In large measure, we have taken over the role played by the process of natural Selection. Individually we must accept our role and take action to preserve our natural heritage. If we make no choice and effort, in essence, we choose denigration, negligence, and finally extinction.

All species face extinction. In the history of life on earth species become extinct eventually. Our greatest dangers come from unrestricted drives to exploit the earth, overpopulation, the acceptance of ignorance, inequalities, and increased state power and inflexibility. If we cannot acknowledge and overcome these conditions, extinction will likely be our fate.

Societies are short lived. There is a huge gap between the traits encoded in DNA that made each individual self-sufficient during the long reign of natural selection and the traits required to form stable, benign, and progressive societies. Among the social insects, ants and bees for instance, social behavior is encoded in DNA. Human societies have no mechanism for the inheritance of successful traits except by cultural control and memes (Dawkins, 1976). Most civilizations have lasted only a few hundred or thousand years and were unprogressive and violent by the time of their demise. How humans and society can reconcile their inherent differences is the ongoing political problem of our time.

Living creatures are made of atoms in various molecular configurations. This is a materialistic state, somewhat like the ideas of the ancient Greek skeptics. No component of nonmaterial existence has ever been found in the physical universe although essences, absolutes, mana, and demons abound in philosophies, myths, and imaginary universes.

Life has the basic purpose to maintain existence and procreate. As conscious and reasoning beings we yearn also for meaning in our lives. As with all things individuals demand, life's meaning can be assured only by individual activity, struggle, and persistence. We share all life's basic purposes but with the beginnings of self-consciousness and reason we have arrived at a new point of mandatory responsibility. We alone can determine the meaning and outcome of life because we alone will determine whether or not life continues on the planet Earth.

Earth's living forms depend on human tolerance because we are now the "Natural Selectors" who, knowingly or not, are now a critical part of the natural selection process. There has been a trend in evolution toward greater complexity, diversity, sensory sensitivity and cooperation. This is the result of chance processes not a goal of evolution. Evolution never had nor has goals.

But with the evolutionary appearance of nature's *Homo sapiens* — the conscious one — there are goals, humanity's goals. With man's present pressure on the environment, diversity is rapidly declining. The ability of living creatures to adapt to the variations of physical environments decreases. Where this ends, if it does, depends on us.

The needs of our time are clear; they are the general goals of freedom from want, slavery, discrimination, ignorance, and full access to the knowledge and benefits of science for all individuals everywhere in the world.

Bibliography

Alberts, B. et al., 1998, *Essential Cell Biology*

Ayers, A. J., 1990, *The Meaning of Life*

Clark, D. and Russell L., 1997, *Molecular Biology*

Clarke, W. R., 1996, *Sex and the Origin of Death*

Chang, R., 1988, *Chemistry*

Cooper, G. M., 1997, *The Cell, A Molecular Approach*

Dawkins, R., 1995, *River Out of Eden*

Dawkins, R., 1976, *The Selfish Gene*

Dennett, D., 1995, *Darwin's Dangerous Idea*

Diamond, J., 2005, *Collapse*

Johnson, P. E., 1991, *Darwin on Trial*

Lockshin, R. et al eds., 1998, *When Cells Die*

Meadows, D. et al., 1972, *Limits to Growth*

——, 1992, *Beyond the Limits*

——, 2004, *Limits to Growth, 30-Year Update*

Raup, D. M., 1991, *Extinction*

Schopf, J. W. ed., 1992, *Major Events in the History of Life*

Schopf, J. W., 1999, *Cradle of Life*

Tainter, J. A., 1988, *Collapse of Complex Societies*

Whitrow, G. J., 1988, *Time in History*

Glossary (Some Facts of Science)

Absolute	St Anselm: "that than which there is nothing greater"
Agnostic	Belief that the ultimate reality is unknown and probably unknowable
Anaerobic	The absence of atmospheric oxygen
Apoptosis	Cell death preprogrammed by DNA
Atomo	Ancient Greek idea of the smallest particle, now the known as the atom
Axiom	Universally accepted principle or rule; self evident truth, used by the ancient Greek Euclid in his *Elements*
Background Radiation	Microwave radiation from all directions of space, temperature of -2.7 degrees. According to Kelvin, this was the first direct evidence of the "Big Bang" event that created the universe from a singularity of infinite density.
Believe	To accept a statement but not necessarily with proof
Bell's Inequality Theorem	In 1964, Bell specified the conditions for locality or non-locality, a way to distinguish between

	the Bohr Copenhagen (CI) and the (EPR), Einstein, Podolsky, and Rosen interpretation of Quantum Mechanics
Big Bang	Gamow's concept that the universe began from an infinitely dense singularity
Binary Fission	Reproduction by division into two essentially equal parts, the typical asexual reproduction process in Prokaryotes (bacteria)
Cepheid Variables	Stars with periodic variations in average light intensity, the period length determines the maximum light output and can be used to find the distance to a star or galaxy once a few Cepheid distances were found by parallax measurements. Discovered by Henrietta Leavitt in 1908; used by Hubble in 1929 to prove the universe expanded
Chloroplast	Symbiotic photosynthetic organelles, originally evolved as separate bacteria cells.
Complementarity	The wave and particle nature of matter are said to "complement" one another — until an observation is attempted, resulting in the collapse into particle or wave-like behavior
Continuous Creation	Hoyle's theory that matter is continuously created in interstellar space, disproved by the microwave background radiation
Continuous Radiation	Energy is a continuous wave, not a particle; disproved by Planck
Correlated pairs	Identical particles moving in opposite directions away from a common source which generated them are said to be "correlated pairs"; they have opposite wave polarizations (vertical or horizontal) if waves, and magnetic spins (up or down), if electrons

Cytoplasm	The cellular material that is within the cell membrane but does not including the organelles.
Deductive Logic	Reasoning from the general to the specific
Deuterium	An isotope of hydrogen, twice as heavy
Diffraction	Wave fronts that pass the edge of an opaque body interfere so that a series of lines can be seen parallel to the edge
Dominant Trait	The one of the two genes (dominant and recessive) of the double helix that is physically evident in an individual creature.
DNA	Deoxyribonucleic acid, the chemical name of the double helix, the molecules that transmit inheritable genetic information
Double Helix	The two molecular strands that make up the inheritable code of living creatures
Doppler Effect	The change in wavelength of light or sound, caused by the relative motions of the transmitting and receiving devices
Electromagnetic Wave	A form of energy carried by electric and magnetic fields oscillating at right angles to the direction of travel at the speed of light, 3×10^{10} cm/sec.
Electron	The particle with a single unit of negative charge
Empirical	Derived from experience or experiment
Entropy	A measure of the degree of disorder and dispersed energy in a system or substance; high entropy is high states of disorder or dispersed energy that can no longer do work
EPR Group	Einstein, Podolsky, and Rosen, prominent scientists who contested Bohr's interpretation of Quantum Theory
Ether	The carrier of light waves, now known to be nonexistent

Eukaryotes	(you-carry-oats) Plants, animals and fungi; DNA enclosed in a nucleus; has many organelles; generally multicellular, bisexual and with larger cells
Evolution	Populations that change with time leading to observable differences among species
Expanding Universe	In 1929, Hubble found the universe was expanding not static; now known to be at an escalating rate
Experimental-Theoretical Science	Facts based on experimental proof are used to create a more general overview, a theory
Frequency	The number of wavelengths per unit of time
Geiger Counter	An instrument to measure radioactivity
Gene	A unit of base pairs (a nucleotide) in DNA that contains the information on how to construction a specific protein
Gravitation	The attractive force between to masses
Half Life	The time required for one half the atoms of a radioactive sample to disintegrate
Hominid	A group including Australopithecus that dates back 4–10 million years, *Homo sapiens* is the only surviving species
Hominoid	A super family including the great apes and man
Heisenberg Uncertainty Principle	Properties of objects, such as position and momentum, cannot be determined with certainty simultaneously
Hypothesis	A potential explanation for a specific set of phenomena
Incompleteness Theorems	Gödel, 1961 (1) A formal system of mathematics will always be incomplete, (2) No set of axioms can prove its own internal consistency

Ideology	A body of doctrine justifying a social or political position
Inductive Logic	Any form of reasoning in which the conclusion is not necessarily supported by the premises
Infer	To derive from a statement unstated but logical conclusions or probabilities
Interference	A wave interaction that can decrease or augment the intensity of waves: light, sound, or water
Inflationary Era	The period of rapid expansion of the universe associated with the Big-Bang Theory
Intuition	Direct perception of a fact independent of observation or reasoning, a feeling
Kilometer	One thousand meters, 0.621 mile
Know	To understand clearly and with certainty a fact by observation and testing
Locality	Describes the notion that entities within our universe must be in some form of direct contact for an interaction to occur
Matter	Mass and energy
Meiosis	Special sexually reproducing cells divide to form sperm and egg cells, each with only one of the paired chromosomes in a somatic or body cell; the form of cell division unique to bisexual species.
Meme	Learned behavior that spreads through a society, cultural replication simulating the inheritable gene
Meson	An elementary particle with a charge and spin of 0 or ½
Metabolism	An energy process leading to growth and maintenance of living cells

Mitochondria	Organelles that produce chemical energy in the form of ATP (adenosine triphosphate) by oxidizing carbohydrates; consume oxygen and release carbon dioxide; they evolved symbiotically from once independent bacteria
Mitosis	The normal cell division process in which the DNA doubles than splits lengthwise forming two identical daughter cells; characteristic of single celled (prokaryote) species and the body cells of multi celled (eukaryote) species
Microwave Background Radiation	This consists of photons coming from every direction in space; the fact that supports the Big Bang theory
Momentum	Mass (m) multiplied by velocity (v)
Muon	A *mu* meson
Mutation	A sudden chemical change from the normal inheritable characteristic in a specific location in DNA
Natural Selection	The genetic and environmental variations that together "select" well adapted creatures for survival and reproduction and reject the poorly adapted
Neutrino	A particle with little or no mass and no electrical charge
Neutron	The particle that carries the positive charge in the atomic nucleus
Non-Euclidian Geometry	Describes elliptic and hyperbolic geometries (curved space) rather than the more familiar Euclidean geometry. The main difference between Euclidean and non-Euclidean geometries is the treatment of parallel lines. In the former they remain equidistant from one another while in the latter initially parallel lines either diverge

	(hyperbolic) or converge and cross one another (elliptic)
Non Local	Describes the concept that entities within our universe do not necessarily have to be in obvious contact with one another and can influence each other's behavior spontaneously
Ontogeny	The sequence of phases from zygote to adult
Organelles	Specialized functional bodies within the cell such as chloroplasts, nuclei, and mitochondria.
Parallax	The apparent displacement of an object when it is viewed from two different positions, for instance, when one eye and then the other is used alone to view an object
Perpetual Motion	The goal of those who want to get something for nothing; it is impossible according to the laws of thermodynamics and conservation of mass.
Photon	The electromagnetic wave that is generated when an electron falls from a high state of energy in to a lower state
Premise	A proposition that supports a conclusion
Probability	The relative frequency that a possible event will actually happen; for instance, flipping a coin heads or tails
Prokaryotes	(pro-carry-oats) Organisms such as bacteria that lack a membrane enclosing the nucleus; DNA floats freely in the cellular cytoplasm; mainly single celled but some form chains or clusters; few organelles
Protista (or Protists)	Single-cell eukaryotes, yeast to amoebas
Quantum	The smallest quantity of radiant energy; Planck's Constant times the frequency of the energy wave

Quantum Mechanics	The science that describes the behavior of atomic and subatomic particles
Quantum pilot wave	In the Bohmian description of quantum mechanics, a "hidden variable," the quantum pilot wave, is introduced to explain experimental observations like those of the two-slit experiment
Quark	Particles from which all atoms are made together with the electron and neutrino
Quark-gluon	A phase that consists of relatively free quarks and gluons which are components matter
Radiometric	To determine time by the radioactive decay of atoms, for instance uranium atoms
Realism	The naïve philosophy that reality is what we observe in our everyday experience
Red Shift	The stretching of light waves causes them to appear longer and redder than when first radiated in a shorter, bluer form. The Doppler Effect, the universal expansion of space, or wave flight through a gravity field, can cause the red shift
Reflect	To return or cast back a light wave from a surface
Refract	The bending of light waves when they move from a medium of one density into a medium of higher or lower density. The effect is due to the differences in velocity of the waves in the two media
Relativistic	Einstein's theory that all motion must be defined relative to a frame of reference and that space and time are relative not absolute concepts
Scientific Method	Systematic knowledge obtained by observation and experimentation of natural phenomena

Singularity	A region of space with infinite density that does not obey the known laws of physics
Snell's Law	The ratio of the difference in angle between an incident ray and a refracted ray on passage through a medium of one density and into a second medium is a constant
Spectrum (Spectra-)	The various wave lengths of light (photons) emitted by an excited atom
Statistical	The science that deals with the classification of a specific property within a group of objects
Supernatural	That which cannot be observed or tested by any means, not natural, with no existence yet can be subjectively thought to exist
Supernova	implosion followed by explosion of a stellar body, produces heavy atoms such as carbon, oxygen, phosphorus, lead
Syllogism	A form of logic producing truth but only if its two premises are true
Symbiosis	A close interaction of two individuals from separate species, which usually benefits both individuals; examples are lichens, chloroplasts and mitochondria.
Thermodynamics, First Law	Mass or energy can be neither created nor destroyed
Thermodynamics, Second Law	Heat cannot be transferred from a colder body to a hotter body without work being done by an outside agent
Trial and Error	Make a test and determine whether the result confirms the expectation
Truth	That which describes and conforms to the real world
Truer Truth	The better truth from a group of possibilities
Uniformatarianism	The argument of the pre-Darwinian geologists championed by C. Lyell that change in Earth's

	topography occurred slowly over vast periods of time not quickly or violently, which was the catastrophists view.
Zygote	A fertilized egg cell, the union of sperm and egg cell nuclei that forms an individual eukaryote